SpringerBriefs in Applied Sciences and Technology

SpringerBriefs present concise summaries of cutting-edge research and practical applications across a wide spectrum of fields. Featuring compact volumes of 50 to 125 pages, the series covers a range of content from professional to academic.

Typical publications can be:

- A timely report of state-of-the art methods
- An introduction to or a manual for the application of mathematical or computer techniques
- A bridge between new research results, as published in journal articles
- A snapshot of a hot or emerging topic
- An in-depth case study
- A presentation of core concepts that students must understand in order to make independent contributions

SpringerBriefs are characterized by fast, global electronic dissemination, standard publishing contracts, standardized manuscript preparation and formatting guidelines, and expedited production schedules.

On the one hand, **SpringerBriefs in Applied Sciences and Technology** are devoted to the publication of fundamentals and applications within the different classical engineering disciplines as well as in interdisciplinary fields that recently emerged between these areas. On the other hand, as the boundary separating fundamental research and applied technology is more and more dissolving, this series is particularly open to trans-disciplinary topics between fundamental science and engineering.

Indexed by EI-Compendex, SCOPUS and Springerlink.

Andriy Nadtochiy · Alla M. Gorb ·
Borys M. Gorelov · Oleksiy Polovina ·
Oleg Korotchenkov

Graphene-Based Polymer Nanocomposites

Models and Applications

Springer

Andriy Nadtochiy
Faculty of Physics
Taras Shevchenko National University
of Kyiv
Kyiv, Ukraine

Borys M. Gorelov
Chuiko Institute of Surface Chemistry
National Academy of Sciences of Ukraine
Kyiv, Ukraine

Oleg Korotchenkov
Faculty of Physics
Taras Shevchenko National University
of Kyiv
Kyiv, Ukraine

Alla M. Gorb
Faculty of Physics
Taras Shevchenko National University
of Kyiv
Kyiv, Ukraine

Oleksiy Polovina
Faculty of Physics
Taras Shevchenko National University
of Kyiv
Kyiv, Ukraine

ISSN 2191-530X   ISSN 2191-5318 (electronic)
SpringerBriefs in Applied Sciences and Technology
ISBN 978-981-97-2791-9   ISBN 978-981-97-2792-6 (eBook)
https://doi.org/10.1007/978-981-97-2792-6

This Springer imprint is published by the registered company Springer Nature Singapore Pte Ltd.
The registered company address is: 152 Beach Road, #21-01/04 Gateway East, Singapore 189721, Singapore

If disposing of this product, please recycle the paper.

*With gratitude to our colleagues and
students, past and present, for a quality
setting that helped us to complete this book*

Preface

Polymer nanocomposites based on graphene and its derivatives are in the focus of numerous theoretical and experimental multidisciplinary studies, which describe novel mechanical, electrical, and thermal properties of the nanocomposites. There has been tremendous attention given to their promise as high-efficiency materials in a wide range of applications including thermal-interface materials, active electrodes for lithium-ion batteries and supercapacitors, membranes for fuel cells, gas separation and proton exchange, materials for electromagnetic shielding and flexible and stretchable electronics. In particular, with the increasing demand for high-elasticity materials used in tires, seals, and shock absorbers, it is plausible to add graphene fillers to matrices for reinforcement.

This book gives a comprehensive introduction to the physics of graphene-based polymer nanocomposites. In addition, particular emphasis is given in this book to the impact of the interfacial interactions on the mechanical and electrical behavior of these materials.

We have attempted to highlight key topics learned and developed by the authors over the past 20 years. We hope this treatment of a family of graphene-based polymer nanocomposites will provide materials scientists, graduate students, engineers, and material designers with a better understanding of this fascinating research and technology field.

Kyiv, Ukraine
January 2024

Andriy Nadtochiy
Alla M. Gorb
Borys M. Gorelov
Oleksiy Polovina
Oleg Korotchenkov

Contents

Acronyms

ABS	Acrylonitrile butadiene styrene
AFM	Atomic force microscopy
ARGET	Activators regenerated by electron transfer
CB	Carbon black
CLD	Cross-linking degree
CMGs	Chemically modified graphenes
CNT	Carbon nanotube
CrGO	Chemically reduced GO
CVD	Chemical vapor deposition
DGEBA	Diglycidyl ether of bisphenol A
DMF	Dimethylformamide
DMG	Dry-milled graphene
FET	Field-effect transistor
FFT	Fast Fourier transform
fGNP	Functionalized GNP
fGO	Functionalized graphene oxide
FLG	Few-layer graphene; a 2D (sheet-like) material, either as a free-standing flake or substrate-bound coating, consisting of a small number (between 2 and about 5) of well-defined, countable, stacked graphene layers of extended lateral dimension
frGO	Functionalized rGO
fTRGO	Functionalized TRGO
GCE	Glassy carbon electrode
GNP	Graphene nanoplatelets, Graphite nanoplate, graphite nanosheet, graphite nanoflake; 2D graphite materials with ABA or ABCA stacking, and having a thickness and/or lateral dimension less than 100 nm
GNS	Graphene nanosheet

GO	Graphene oxide; the oxidized analogy of graphene, widely recognized as the only intermediate or precursor for obtaining graphenes on a large scale
GQD	Graphene quantum dot
HDPE	High-density polyethylene
HPC-Py	Pyrene-containing hydroxypropyl cellulose
IPLs	Interphase layers
iPP	Isotactic polypropylene
ITO	Indium-tin oxide
LB	Langmuir–Blodgett
LbL	Layer-by-Layer
LLDPE	Linear low-density polyethylene
MG	Methylene green
MLG	Multilayer graphene; same as FLG with greater layer numbers (typically up to 10)
MSA	Multiple scattering approach
MWCNT	Multi-walled carbon nanotube
NR	Natural rubber
N-TRGO	Nitrogen-TRGO
P3HT	Poly(3-hexylthiophene)
PA11	Polyamide 11
PA12	Polyamide 12
PA6	Polyamide 6
PAA	Poly(acrylic acid)
PAH	Polycyclic aromatic hydrocarbons
PANI	Polyaniline
PB	Polybutadiene
PBA	Poly(butyl acrylate)
PC	Polycarbonate
PCL	Polycaprolactone
PDDA	Poly(diallyldimethylammonium chloride)
PDMS	Poly(dimethylsiloxane)
PE	Polyethylene
PEG	Poly(ethyl glycol)
PEI	Polyethylenimine
PET	Polyethylene terephthalate
pG	Pristine graphene
PHPMA	Poly(N-(2-hydroxyphenyl)methacrylamide)
PI	Polyimide
PIL	Poly(ionic liquid)
PLA	Polylactic acid
PMMA	Poly(methyl methacrylate)
PNCs	Polymer nanocomposites
PNIPAM	Poly(N-isopropylacrylamide)
POSS	Polyhedral oligomeric silsesquioxane

PP	Polypropylene
PPESO$_3^-$	Poly(2,5-bis(3-sulfonatopropoxy)-1,4-ethynylphenylene-alt-1,4-ethynylphenylene)
PS	Polystyrene
PSS	Poly(styrenesulfonate)
PSSA-g-PPY	Poly(styrenesulfonic acid-g-pyrrole)
PTi	Polythiophene
PU	Polyurethane
PVA	Poly(vinyl acetate)
PVC	Poly(vinyl chloride)
QP$_4$VP-co-PCN	Cationic azo polyelectrolyte
RAFT	Reversible addition-fragmentation chain transfer
rGO	Reduced graphene oxide
RVE	Representative volume element
SAN	Styrene-acrylonitrile
SANS	Small-angle neutron scattering
SCMC	Sodium carboxymethyl cellulose
SDBS	Sodium dodecyl benzene sulfonate
SLS	Sodium lignosulfonate
SPANI	Sulfonated polyaniline
sPS	Syndiotactic polystyrene
TCF	Transparent conducting film
TGA	Thermogravimetric analysis
TMDSC	Temperature-modulated differential scanning calorimetry
TPU	Thermoplastic polyurethane
TRGO	Thermally reduced graphene oxide
UHMWPE	Ultra-high-molecular-weight polyethylene
UPS	Ultrasound phase spectroscopy

Symbols

$\mathbf{C}$	Overall elastic stiffness tensor of the composite
$\mathbf{C}^i$	Stiffness tensor for the interface
$\mathbf{C}^m$	Stiffness tensors of the matrix
$\mathbf{C}^p$	Stiffness tensors of the effective particle
c^i	Volume fraction for the interface
C_{ijkl}	Components of the linear-elastic stiffness tensor
c^m	Effective matrix volume fraction
c^p	Effective particle volume fraction
d	Diameter of the filler particle
D_i	i-th component of the electric displacement vector
$D(t)$	Creep compliance
E	Elastic or Young's modulus
E_i	i-th component of the electric field strength
$E(t)$	Relaxation modulus of the polymer
f	Sound frequency
f_c	Volume fraction of filler at percolation threshold
f_{fil}	Volume fraction of the filler particle
f_{int}	Volume fraction of the interphase
h	Thickness of the filler particle
$\mathbf{I}$	Identity tensor
L	Length of the fiber-like (cylinder) filler particle
n_f	Length aspect ratio of the fiber-like particle (the dominant dimension over the minor dimension of the filler)
n_p	Length aspect ratio of the platelet
q	Externally applied body force
R_d	Radius of the disk-shaped filler nanoplatelet
R_s	Radius of the spherical filler particle
S_{fil}	Surface area of the filler particle
S_{kl}	Components of the strain tensor
$\mathbf{S}^p$	Dilute strain-concentration tensor of the effective particles

$\mathbf{S}^{pi}$ Dilute strain-concentration tensors of the effective interface formed by filler inclusion in a matrix

t Interphase layer thickness

T_{ij} Components of the stress tensor

u Mechanical displacement

v, v_0 Sound velocity

V_{fil} Volume of the filler particle

V_{int} Volume of the interphase layer

V_{RVE} Volume of the RVE

η Viscosity

$\mathfrak{L}$ Characteristic filler length

λ Lame constant

μ Scaling exponent determining the power of the conductivity increase above the percolation threshold f_c; Lame constant

v Poisson ratio

ρ Density of a material

σ Electrical conductivity

σ_0 Conductivity of the filler particles

φ Electric potential

Ξ^p Eshelby tensor

ω Angular frequency

Part I
State of the Art, Mechanical Properties

Chapter 1
Introduction

Extensive studies have been made to produce novel materials with unique physical and chemical properties by combining matrix and filler nanomaterials. These high-performance composite materials have been widely employed in multifaceted applications ranging from electrical devices to biological appliances [1]. However, preparing homogeneous composite materials with nanosheets such as graphene is still very challenging due to the strong Van der Waals attraction between the nanosheets [2].

A vast amount of studies have revealed that graphene-based composites provide higher performance compared with the individual components under the same condition because of the synergistic effects between each component and simultaneous creation of a unique filler network in the matrix [3–6]. Consequently, the composites are greatly appreciated for designing a new type of versatile materials for high-performance transparent conducting films (TCFs), field-effect transistors (FETs), energy storage and generation systems (lithium-ion batteries, supercapacitors, and solar cells), sensors, communication networks, electrodes, bioimaging, drug delivery systems, gas barriers, photocatalysts, etc. However, graphene-based composites are typically disordered, worsening their reproducibility and device performance.

As an emerging electronic material, graphene is considered to be the thinnest functional nanomaterial for thin-layer and multilayer structures in advanced electronic devices. In particular, the assembly of graphene sheets into a layer-stacking macroscopic structure delivers excellent mechanical strength and reliability, optical transparency, electrical/thermal conductivity, and current-carrying capacity [7–10]. These high-quality multilayer structures are crucial in various applications, including transparent electrodes, low-resistance wiring, and heat spreaders [11]. They were fabricated on arbitrary substrates using transfer techniques and chemical vapor deposition (CVD) [12–17], vacuum-assisted filtration [18, 19], the Langmuir–Blodgett (LB) assembly, in which monolayers are transferred from a liquid/gas interface to a solid surface [20–23], and Layer-by-Layer (LbL) assembly [24–28]. The fabrication

A. Nadtochiy et al., *Graphene-Based Polymer Nanocomposites*,
SpringerBriefs in Applied Sciences and Technology,
https://doi.org/10.1007/978-981-97-2792-6_1

of graphene films on various substrates was also performed using spin- or spray-coating [29], drop-casting [30], dip-coating [31], electrophoretic deposition [42]. Among these fabrication methods, LB and LbL techniques are advantageous because they precisely control the film thickness and architecture at the molecular level. For example, films with a high degree of molecular organization can be obtained, and the spatial arrangement of the film material can be investigated by the LB technique. The LbL method is superior due to the simplicity of the film deposition process for a wide variety of materials.

In the LB process, an insoluble monolayer (which is called Langmuir film) of amphiphile molecules is first spread on the surface of a water phase (referred to as the subphase). After compressing the monolayer into a highly condensed state, it is transferred onto solid support, resulting in layer-by-layer growth. The LbL technique is achieved through the sequential adsorption of oppositely charged components by attractive forces such as electrostatic interactions, and hydrogen bonding.

Polymeric materials have been widely used in LbL assembly due to their functional groups for electrostatic interactions or hydrogen bonding [2]. Usually, as-prepared graphene oxide (GO) is negatively charged in an aqueous solution which originates from its carboxylic acid and phenolic hydroxyl functional groups. Then, positively charged polyelectrolytes have been selected for the counter component in graphene–polymer LbL assembly [32–37]. Furthermore, hydrogen bonding can act as an attractive force between GO and the polymer due to abundant oxygen-functional groups in GO [38, 39]. Weak Van der Waals forces can also be appreciable in LbL assembly between GO and biopolymers [40, 41]. Due to the outstanding mechanical properties of graphene, graphene–polymer nanocomposites have been widely explored by using graphene derivatives as nanofillers. Polymers were also used as monolayer components in the LB method since superior mechanical and thermal stabilities are expected for their Langmuir monolayers [43]. LB-assembled polymers were successfully combined with graphene and graphene derivatives [44–46]. Some noteworthy examples of LbL assembly between graphene and polymeric materials are summarized in Table 1.1.

The research on polymeric nanocomposites during the last decades in part has been focused on the development of conventional nano-filled polymers aimed to obtain nanocomposites with improved engineered functional properties. Nanosized inorganic or organic powders are combined with polymers to form polymer nanocomposites whose physical properties and performance significantly differ from those of the component materials. These composites have high thermal and mechanical stability, multifunctional and chemical functionalization capabilities, and large interphase regions. Usually, the nanocomposites exhibit improved properties such as low density, high mechanical stiffness, strength, toughness, anti-corrosion ability, and thermal insulation.

It is appropriate to mention that most graphene–polymer nanocomposites do not exclusively contain individual pristine graphene sheets dispersed in a polymer matrix, but rather a mixture of individual graphene sheets (sometimes functionalized) and few-layer graphene stacks. There are numerous ways to incorporate nanofillers into a polymer matrix such as solution processing, *in situ* polymerization, melt blending,

Table 1.1 Samples of graphene/polymer hybrids based on LbL assembly [2]

(+) Component	(−) Component	Substrate	Applications	References
PANI	GO	ITO/glass	Supercapacitor	[47]
PANI	GO	Quartz	Electrode	[48]
PANI	rGO	Stainless steel	Supercapacitor	[49]
PANI	PSS-GO	ITO/glass	Sensor	[50]
PDDA	GO	ITO/glass	Electrode	[51]
PDDA	GO	GCE	Electrocatalyst	[52]
PDDA	GO	Nafion	Membrane	[53]
PEI	PAA-rGO	GCE	Biosensing	[54]
PEI	GO	PET	Gas barrier	[33, 34]
PVA	GO	Quartz	–	[35]
PVA	GO	Si wafer	–	[39]
PAH	GO	$CaCO_3$	Drug encapsulation	[36]
Poly-L-lysine	Heparin-rGO	PCL scaffold	Neural scaffolds	[55]
QP_4VP-co-PCN	rGO	ITO/glass	Supercapacitor	[37]
Chitosan-rGO-enzyme	PSS	Au/Quartz	Biosensing	[56]

latex technology, and other commonly used methods [57]. Several types of graphene produced from graphite are frequently used as filler particles. These include (i) the oxidation of graphite and subsequent chemical reduction of graphite oxide in the presence of PSS, (ii) the oxidation of graphite and subsequent thermal reduction of graphite oxide, and (iii) liquid-phase exfoliation of as-produced graphite by bath sonication in the presence of surfactant (sodium cholate) [57]. In general, these three methods are suitable for large-scale graphene production required for industrial polymer nanocomposite applications. Starting from graphite, or its derivatives, and using a solvent-based processing method offers significant economic advantages over methods such as mechanical exfoliation, chemical vapor deposition, and epitaxial growth.

It has been well established that the physical properties of the nanocomposites crucially depend on the interface characteristics. However, despite significant insights into the nature of graphene–polymer interactions at the interface have been achieved, there is yet a lack of complete understanding of many interface and interphase issues in the nanocomposites [58]. Further investigations into the relationship between composite properties and interfacial optimization are needed to achieve optimal performance of polymer matrices with graphene sheet species.

Due to their novel multifunctional properties, polymer nanocomposites can be used in a broad range of applications such as outer space components, automobiles, coatings, adhesives, packaging materials, microelectronic packaging, drug delivery devices, sensors, and membranes [1, 59].

The book is aimed at wide audiences to provide a comprehensive discussion of the physics behind the rich variety of graphene-based polymer nanocomposites. Both experimental results and model descriptions of the relationships between the effective properties of the nanocomposites and the properties of interphase and interfacial regions are discussed. The introduction chapter provides a very brief overview of polymer–graphene composites focusing on the thin-layer and multilayer structures as well as conventional nano-filled polymers for engineered functional applications.

Chapter 2 gives a brief description of the basic knowledge of graphene and its derivatives including graphene oxide, reduced graphene, graphene nanoplatelets, and quantum dots. Their fabrication and processing directed toward observable properties are particularly taken into account.

Chapter 3 provides state-of-the-art multifunctional graphene-based polymer nanocomposite materials, including morphological details and percolation aspects with emphasis given to models of interphase and interfacial interactions.

Chapter 4 focuses on the static and dynamic mechanics of polymer–graphene composites and the modeling of interphase effects in the overall mechanical behavior of these materials. This chapter also introduces the sonic wave assisted analysis of graphene-based composites.

References

1. N. Karak (ed.), *Nanomaterials and Polymer Nanocomposites: Raw Materials to Applications* (Elsevier, Amsterdam, 2019)
2. T. Lee, S.H. Min, M. Gu, Y.K. Jung, W. Lee, J.U. Lee, D.G. Seong, B.-S. Kim, Layer-by-layer assembly for graphene-based multilayer nanocomposites: Synthesis and applications. Langmuir **27**, 3785–3796 (2015)
3. U. Szeluga, B. Kumanek, B. Trzebicka, Synergy in hybrid polymer/nanocarbon composites. A review. Compos. Part A Appl. Sci. Manuf. **73**, 204–231 (2015)
4. Y.-M. Jen, J.-C. Huang, K.-Y. Zheng, Synergistic effect of multi-walled carbon nanotubes and graphene nanoplatelets on the monotonic and fatigue properties of uncracked and cracked epoxy composites. Polymers **12**, 1895 (2020)
5. I. Charitos, G. Georgousis, P.A. Klonos, A. Kyritsis, D. Mouzakis, Y. Raptis, A. Kontos, E. Kontou, The synergistic effect on the thermomechanical and electrical properties of carbonaceous hybrid polymer nanocomposites. Polym. Test. **95**, 107102 (2021)
6. A. Moradi, R. Ansari, M.K. Hassanzadeh-Aghdam, Synergistic effect of carbon nanotube/graphene nanoplatelet hybrids on the elastic and viscoelastic properties of polymer nanocomposites: finite element micromechanical modeling. Acta Mech. (2023). https://doi.org/10.1007/s00707-023-03782-1
7. K.S. Kim, Y. Zhao, H. Jang, S.Y. Lee, J.M. Kim, K.S. Kim, J.-H. Ahn, P. Kim, J.-Y. Choi, B.H. Hong, Large-scale pattern growth of graphene films for stretchable transparent electrodes. Nature **457**, 706–710 (2009)
8. R. Murali, Y. Yang, K. Brenner, T. Beck, J.D. Meindl, Breakdown current density of graphene nanoribbons. Appl. Phys. Lett. **94**, 243114 (2009)
9. S. Biswas, L.T. Drzal, Multilayered nano-architecture of variable sized graphene nanosheets for enhanced supercapacitor electrode performance. ACS Appl. Mater. Interfaces **2**, 2293–2300 (2010)
10. A.A. Balandin, Thermal properties of graphene and nanostructured carbon materials. Nat. Mater. **10**, 569–581 (2011)

11. H. Murata, Y. Nakajima, N. Saitoh, N. Yoshizawa, T. Suemasu, K. Toko, High-electrical-conductivity multilayer graphene formed by layer exchange with controlled thickness and interlayer. Sci. Rep. **9**, 4068 (2019)
12. K.S. Novoselov, A.K. Geim, S.V. Morozov, D. Jiang, Y. Zhang, S.V. Dubonos, I.V. Grigorieva, A.A. Firsov, Electric field effect in atomically thin carbon films. Science **306**, 666–669 (2004)
13. S. Unarunotai, J.C. Koepke, C.-L. Tsai, F. Du, C.E. Chialvo, Y. Murata, R. Haasch, I. Petrov, N. Mason, M. Shim, J. Lyding, J.A. Rogers, Layer-by-layer transfer of multiple, large area sheets of graphene grown in multilayer stacks on a single SiC wafer. ACS Nano **4**, 5591–5598 (2010)
14. C.-Y. Su, A.-Y. Lu, C.-Y. Wu, Y.-T. Li, K.-K. Liu, W. Zhang, S.-Y. Lin, Z.-Y. Juang, Y.-L. Zhong, F.-R. Chen, L.-J. Li, Direct formation of wafer scale graphene thin layers on insulating substrates by chemical vapor deposition. Nano Lett. **11**, 3612–3616 (2011)
15. J.-J. Chen, J. Meng, Y.-B. Zhou, H.-C. Wu, Y.-Q. Bie, Z.-M. Liao, D.-P. Yu, Layer-by-layer assembly of vertically conducting graphene devices. Nat. Commun. **4**, 1921 (2013)
16. K. Murakami, S. Tanaka, A. Hirukawa, T. Hiyama, T. Kuwajima, E. Kano, M. Takeguchi, J.-I. Fujita, Direct synthesis of large area graphene on insulating substrate by gallium vapor-assisted chemical vapor deposition. Appl. Phys. Lett. **106**, 093112 (2015)
17. Ueno, K., Ichikawa, H., Uchida, T.: Effect of current stress during thermal CVD of multilayer graphene on cobalt catalytic layer. Jpn. J. Appl. Phys. **55**, 04EC13 (2016)
18. X. Xie, L. Qu, C. Zhou, Y. Li, J. Zhu, H. Bai, G. Shi, L. Dai, An asymmetrically surface-modified graphene film electrochemical actuator. ACS Nano **4**, 6050–6054 (2010)
19. Q. Liang, X. Yao, W. Wang, Y. Liu, C.P. Wong, A three-dimensional vertically aligned function-alized multilayer graphene architecture: An approach for graphene-based thermal interfacial materials. ACS Nano **5**, 2392–2401 (2011)
20. L.J. Cote, F. Kim, J. Huang, Langmuir-Blodgett assembly of graphite oxide single layers. J. Am. Chem. Soc. **131**, 1043–1049 (2009)
21. Q. Zheng, W.H. Ip, X. Lin, N. Yousefi, K.K. Yeung, Z. Li, J.-K. Kim, Transparent conductive films consisting of ultralarge graphene sheets produced by Langmuir-Blodgett assembly. ACS Nano **5**, 6039–6051 (2011)
22. S. Sharma, K. Qanungo, A review on graphene in energy devices. AIP Conf. Proc. **2535**, 030012 (2023)
23. J.M. Devida, F. Herrera, M.A.D. Millone, F.G. Requejo, D. Pallarola, Electrochemical fine-tuning of the chemoresponsiveness of Langmuir-Blodgett graphene oxide films. ACS Omega **8**, 27566–27575 (2023)
24. N.A. Kotov, I. Dékány, J.H. Fendler, Ultrathin graphite oxide-polyelectrolyte composites prepared by self-assembly: Transition between conductive and non-conductive states. Adv. Mater. **8**, 637–641 (1996)
25. M. Yang, Y. Hou, N.A. Kotov, Graphene-based multilayers: Critical evaluation of materials assembly techniques. Nano Today **7**, 430–447 (2012)
26. K. Ariga, Y. Yamauchi, G. Rydzek, Q. Ji, Y. Yonamine, K.C.W. Wu, J.P. Hill, Layer-by-Layer nanoarchitectonics: Invention, innovation, and evolution. Chem. Lett. **43**, 36–68 (2014)
27. K. Hu, D.D. Kulkarni, I. Choi, V.V. Tsukruk, Graphene-polymer nanocomposites for structural and functional applications. Prog. Polym. Sci. **39**, 1934–1972 (2014)
28. R.F. de Oliveira, A. de Barros, M. Ferreira, Nanostructured films: Langmuir–Blodgett (LB) and Layer-by-Layer (LbL) techniques, in *Nanostructures*, eds. by A.L. Da Róz, M. Ferreira, F. de Lima Leite, Jr., O.N. Oliveira (Elsevier, Amsterdam, 2017), pp. 105–123
29. H.C. Schniepp, J.-L. Li, M.J. McAllister, H. Sai, M. Herrera-Alonso, D.H. Adamson, R.K. Prud'homme, R. Car, D.A. Saville, I.A. Aksay, Functionalized single graphene sheets derived from splitting graphite oxide. J. Phys. Chem. B **110**, 8535–8539 (2006)
30. C. Gómez-Navarro, R.T. Weitz, A.M. Bittner, M. Scolari, A. Mews, M. Burghard, K. Kern, Electronic transport properties of individual chemically reduced graphene oxide sheets. Nano Lett. **7**, 3499–3503 (2007)
31. X. Wang, L. Zhi, K. Müllen, Transparent, conductive graphene electrodes for dye-sensitized solar cells. Nano Lett. **8**, 323–327 (2008)

32. N.I. Kovtyukhova, P.J. Ollivier, B.R. Martin, T.E. Mallouk, S.A. Chizhik, E.V. Buzaneva, A.D. Gorchinskiy, Layer-by-layer assembly of ultrathin composite films from micron-sized graphite oxide sheets and polycations. Chem. Mater. **11**, 771–778 (1999)
33. Y.H. Yang, L. Bolling, M.A. Priolo, J.C. Grunlan, Super gas barrier and selectivity of graphene oxide-polymer multilayer thin films. Adv. Mater. **25**, 503–508 (2013)
34. J.T. Chen, Y.J. Fu, Q.F. An, S.C. Lo, S.H. Huang, W.S. Hung, C.C. Hu, K.R. Lee, J.Y. Lai, Tuning nanostructure of graphene oxide/polyelectrolyte LbL assemblies by controlling pH of GO suspension to fabricate transparent and super gas barrier films. Nanoscale **5**, 9081–9088 (2013)
35. D. Chen, X. Wang, T. Liu, X. Wang, J. Li, Electrically conductive poly(vinyl alcohol) hybrid films containing graphene and layered double hydroxide fabricated via layer-by-layer self-assembly. ACS Appl. Mater. Interfaces **2**, 2005–2011 (2010)
36. R. Kurapati, A.M. Raichur, Graphene oxide based multilayer capsules with unique permeability properties: facile encapsulation of multiple drugs. Chem. Commun. **48**, 6013–6015 (2012)
37. D. Wang, X. Wang, Self-assembled graphene/azo polyelectrolyte multilayer film and its application in electrochemical energy storage device. Langmuir **27**, 2007–2013 (2011)
38. J. Zhu, H. Zhang, N.A. Kotov, Thermodynamic and structural insights into nanocomposites engineering by comparing two materials assembly techniques for graphene. ACS Nano **7**, 4818–4829 (2013)
39. X. Zhao, Q. Zhang, Y. Hao, Y. Li, Y. Fang, D. Chen, Alternate multilayer films of poly(vinyl alcohol) and exfoliated graphene oxide fabricated via a facial layer-by-layer assembly. Macro-molecules **43**, 9411–9416 (2010)
40. K. Hu, M.K. Gupta, D.D. Kulkarni, V.V. Tsukruk, Ultra-robust graphene oxide-silk fibroin nanocomposite membranes. Adv. Mater. **25**, 2301–2307 (2013)
41. K. Hu, L.S. Tolentino, D.D. Kulkarni, C. Ye, S. Kumar, V.V. Tsukruk, Written-in conductive patterns on robust graphene oxide biopaper by electrochemical microstamping. Angew. Chem. Int. Ed. **52**, 13784–13788 (2013)
42. L. Chen, Y. Tang, K. Wang, C. Liu, S. Luo, Direct electrodeposition of reduced graphene oxide on glassy carbon electrode and its electrochemical application. Electrochem. Commun. **13**, 133–137 (2011)
43. M. Aminuzzaman, M. Mitsuishi, T. Miyashita, Fabrication of fluorinated polymer nanosheets using the Langmuir-Blodgett technique: characterization of their surface properties and applications. Polym. Int. **59**, 583–596 (2010)
44. T.-H. Kim, H. Kim, K.-I. Choi, J. Yoo, Y.-S. Seo, J.-S. Lee, J. Koo, Graphene oxide monolayer as a compatibilizer at the polymer-polymer interface for stabilizing polymer bilayer films against dewetting. Langmuir **32**, 12741–12748 (2016)
45. R.A. Soler-Crespo, L. Mao, J. Wen, H.T. Nguyen, X. Zhang, X. Wei, J. Huang, S.T. Nguyen, H.D. Espinosa, Atomically thin polymer layer enhances toughness of graphene oxide mono-layers. Matter **1**, 369–388 (2019)
46. O.N. Oliveira Jr., L. Caseli, K. Ariga, The past and the future of Langmuir and Langmuir-Blodgett films. Chem. Rev. **122**, 6459–6513 (2022)
47. T. Lee, T. Yun, B. Park, B. Sharma, H.K. Song, B.S. Kim, Hybrid multilayer thin film supercapacitor of graphene nanosheets with polyaniline: Importance of establishing intimate electronic contact through nanoscale blending. J. Mater. Chem. **22**, 21092–21099 (2012)
48. K. Sheng, H. Bai, Y. Sun, C. Li, G. Shi, Layer-by-layer assembly of graphene/polyaniline multilayer films and their application for electrochromic devices. Polymer **52**, 5567–5572 (2011)
49. Z. Gao, W. Yang, J. Wang, H. Yan, H. Yao, J. Ma, B. Wang, M. Zhang, L. Liu, Electrochemical synthesis of layer-by-layer reduced graphene oxide sheets/polyaniline nanofibers composite and its electrochemical performance. Electrochim. Acta **91**, 185–194 (2013)
50. J. Luo, Y. Chen, Q. Ma, R. Liu, X. Liu, Layer-by-layer self-assembled hybrid multilayer films based on poly(sodium 4-styrenesulfonate) stabilized graphene with polyaniline and their electrochemical sensing properties. RSC Adv. **3**, 17866–17873 (2013)

51. X. Xu, D. Huang, K. Cao, M. Wang, S.M. Zakeeruddin, M. Grätzel, Electrochemically reduced graphene oxide multilayer films as efficient counter electrode for dye-sensitized solar cells. Sci. Rep. **3**, 1489 (2013)
52. D. Huang, B. Zhang, Y. Zhang, F. Zhan, X. Xu, Y. Shen, M. Wang, Electrochemically reduced graphene oxide multilayer films as metal-free electrocatalysts for oxygen reduction. J. Mater. Chem. A **1**, 1415–1420 (2013)
53. T. Yuan, L. Pu, Q. Huang, H. Zhang, X. Li, H. Yang, An effective methanol-blocking membrane modified with graphene oxide nanosheets for passive direct methanol fuel cells. Electrochim. Acta **117**, 393–397 (2014)
54. R. Pei, X. Cui, X. Yang, E. Wang, Assembly of alternating polycation and DNA multilayer films by electrostatic layer-by-layer adsorption. Biomacromolecules **2**, 463–468 (2001)
55. K. Zhou, G.A. Thouas, C.C. Bernard, D.R. Nisbet, D.I. Finkelstein, D. Li, J.S. Forsythe, Method to impart electro- and biofunctionality to neural scaffolds using graphene-polyelectrolyte multilayers. ACS Appl. Mater. Interfaces **4**, 4524–4531 (2012)
56. M.M. Barsan, M. David, M. Florescu, L. Țugulea, C.M.A. Brett, A new self-assembled layer-by-layer glucose biosensor based on chitosan biopolymer entrapped enzyme with nitrogen doped graphene. Bioelectrochemistry **99**, 46–52 (2014)
57. N. Grossiord, M.-C. Hermant, E. Tkalya, Electrically conductive polymer-graphene composites prepared using Latex technology, in *Polymer-Graphene Nanocomposites* ed. by V. Mittal (RSC, Cambridge, 2012), pp. 66–85
58. A.N. Netravali, K.L. Mittal (eds.), *Interface/Interphase in Polymer Nanocomposites* (Wiley, Hoboken, NJ, 2017)
59. V. Mittal (ed.), *Polymer-Graphene Nanocomposites* (RSC, Cambridge, 2012)

Chapter 2
Graphene and Its Derivatives: Concise Review of Some Basic Fundamentals

Graphene has a wide variety of effective properties such as high carrier mobility, optical transparency, and exceptional electrical and thermal conductivity. Considerable interest in graphene has been motivated by its huge potential to advance several technologically important areas including materials science, biomedical science, and energy [1, 2]. Despite great promise and potential, the graphene-related field holds many challenges that hamper the rapid development of graphene-based functional devices. In particular, pristine graphene (pG) has poor solubility, low reactivity, and intrinsic zero bandgap energy. Production of high-quality pG in large quantities remains a great challenge. Many of these challenges can be overcome with the chemical functionalization of the material, which can open pG bandgap, tune conductivity, improve solubility, and enhance the properties of graphene-based composite materials.

2.1 Morphology and Structural Disorders

Graphene is a single-atom-thick sheet of sp^2-hybridized carbon atoms tightly packed in a two-dimensional (2D) honeycomb lattice (see Fig. 2.1) with a carbon–carbon distance of 1.42 Å [4, 5]. Pristine graphene is considered a unique zero bandgap semiconductor because its conduction band and valence bands encounter at the Dirac points.

The honeycomb structure can be seen as a triangular lattice with a basis of two atoms per unit cell. The lattice vectors can be written as [6]:

$$\mathbf{a_1} = \frac{a}{2}(3, \sqrt{3}), \mathbf{a_2} = \frac{a}{2}(3, -\sqrt{3}), \tag{2.1}$$

© The Author(s), under exclusive license to Springer Nature Singapore Pte Ltd. 2024   11
A. Nadtochiy et al., *Graphene-Based Polymer Nanocomposites*,
SpringerBriefs in Applied Sciences and Technology,
https://doi.org/10.1007/978-981-97-2792-6_2

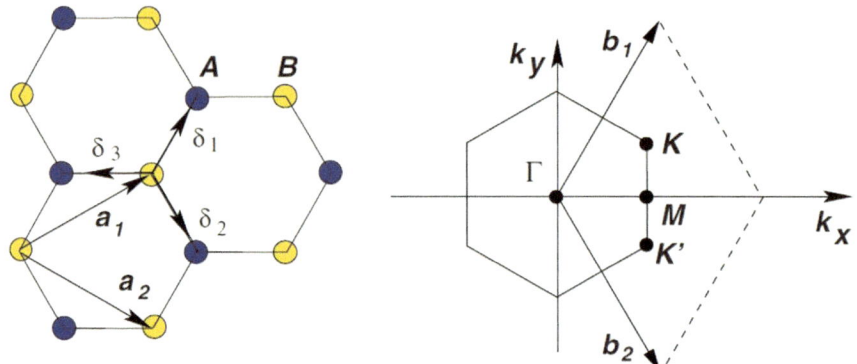

Fig. 2.1 Lattice structure of graphene (left), made out of two interpenetrating triangular lattices ($\mathbf{a}_1$ and $\mathbf{a}_2$ are the lattice unit vectors, and δ_i, $i = 1, 2, 3$ are the nearest-neighbor vectors) and corresponding Brillouin zone (right) (reused with permissions from [6] Copyright ©2009, American Physical Society)

where $a \approx 1.42$ Å is the carbon–carbon distance. The reciprocal-lattice vectors are given by

$$\mathbf{b_1} = \frac{2\pi}{3a}(1, \sqrt{3}), \mathbf{b_2} = \frac{2\pi}{3a}(1, -\sqrt{3}), \tag{2.2}$$

Remember that carbon is the sixth element in the periodic table. As seen in Fig. 2.2a, the nucleus of a carbon atom is surrounded by six electrons, four of which are valence electrons. A ground-state electronic configuration of carbon atom is $1s^2 2s^2 2P_x^1 2P_y^1 2P_z^0$, as shown in Fig. 2.2b. For convenience, the energy level of $2p_z$ is kept with no electron, though it is equivalent to the energy levels of $2p_x$ and $2p_y$. These electrons in the valence shell of carbon can form three types of hybridization, namely sp, sp^2 and sp^3. Figure 2.2c illustrates the formation of sp^2 hybrids. When carbon atoms share sp^2 electrons with their three neighboring carbon atoms, they form a layer of honeycomb network of planar structure, which is also called monolayer graphene.

In a typical sp^2 hybridization of two neighboring carbon atoms on the graphene layer (see Fig. 2.2e), an out-of-plane π bond is made up by $2p_z$ orbitals which are perpendicular to the planar structure. In contrast, an in-plane σ bond is formed by the sp^2 ($2s$, $2p_x$ and $2p_y$) hybridized orbitals. The resulting covalent σ bond is even stronger than the sp^3 hybridized carbon–carbon bonds in diamonds, thus yielding remarkable mechanical properties of the monolayer graphene.

Graphene also has remarkable electronic properties. Because of the robustness and specificity of the σ bonding, it is very hard for alien atoms to replace the carbon atoms in the honeycomb lattice. This is one of the reasons why the electron mean free path in graphene can be so long, frequently reaching values of up to 1 μm.

Of particular importance for the physics of graphene are the two Dirac points, $\mathbf{K}$ and $\mathbf{K}'$, at the corners of the graphene Brillouin zone (BZ) (right picture in Fig. 2.1).

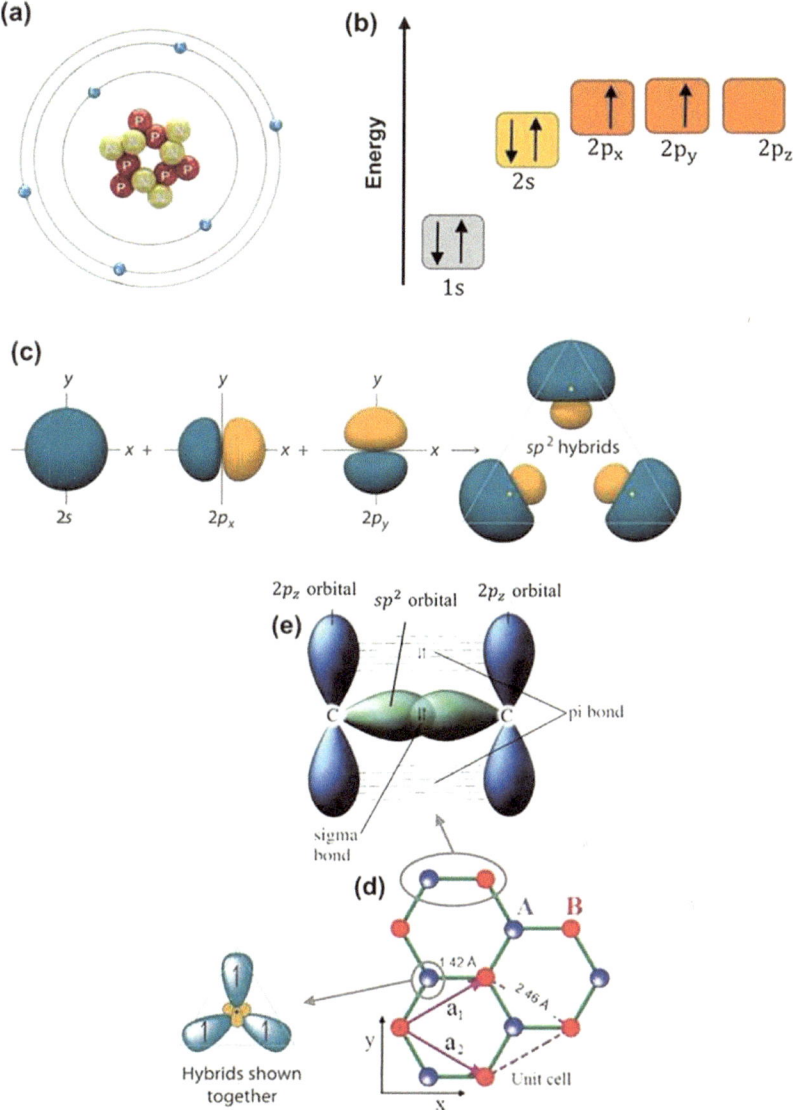

Fig. 2.2 a Atomic structure of a carbon atom. **b** Energy levels of outer electrons in carbon atoms. **c** The formation of sp^2 hybrids. **d** The crystal lattice of graphene, where **a** and **b** are carbon atoms belonging to different sub-lattices, $\mathbf{a_1}$ and $\mathbf{a_2}$ are unit-cell vectors (see also Fig. 2.1). **e** Sigma and pi bonds formed by sp^2 hybridization (reused with permissions from [7] Copyright ©2018 The Author(s). Published by National Institute for Materials Science in partnership with Taylor & Francis Group)

Positions of the Dirac points in momentum space are given by

$$\mathbf{K} = (\frac{2\pi}{3a}, \frac{2\pi}{3\sqrt{3}a}), \mathbf{K}' = (\frac{2\pi}{3a}, -\frac{2\pi}{3\sqrt{3}a}). \tag{2.3}$$

However, one needs to keep in mind that, in the standard harmonic approximation [8], the long-range order of two-dimensional lattices should be destroyed by thermal fluctuation, so perfectly flat graphene does usually not exist [9, 10]. Therefore, graphene is not immune to disorder, and its physical properties are controlled by extrinsic as well as intrinsic disordering effects that are unique to this system. The disorder can be introduced into the crystal structure of graphene during the synthesis process. Among the intrinsic sources of disorder, highlighted are surface ripples (or corrugations) and topological defects. Extrinsic disorder can come about in many different forms such as adatoms, vacancies, sp^3 defects, charges on top of graphene or in the substrate, and extended defects such as cracks and edges. All these kinds of disorders are crucially important in controlling the graphene properties.

There are two types of regular edges in a honeycomb structure, zigzag and armchair. These are exemplified in Fig. 2.3 showing a honeycomb lattice having zigzag edges along the x direction and armchair edges along the y direction. Choosing the ribbon to be infinite in the x direction produces a graphene nanoribbon with zigzag edges. Conversely, if the ribbon is macroscopically large along the y but finite in the x direction, this yields a graphene nanoribbon with armchair edges.

Fig. 2.3 A piece of a honeycomb lattice displaying both zigzag and armchair edges (reused with permissions from [6] Copyright ©2009, American Physical Society)

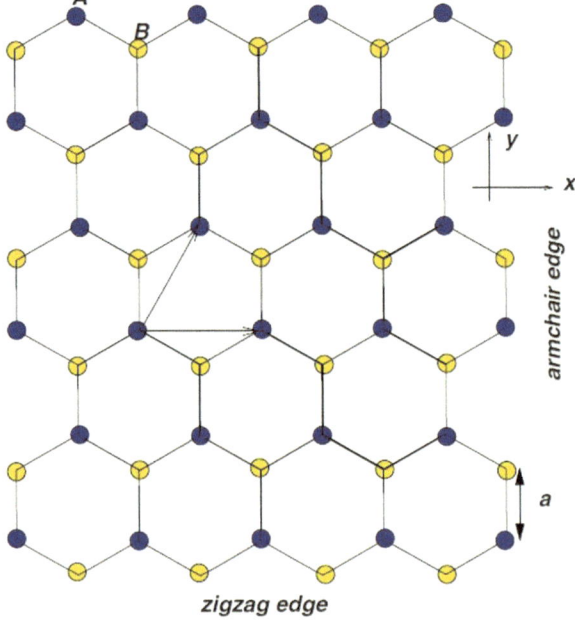

The resulting physical and chemical properties are directly related to the edge topology [11]. In that respect, a zigzag periphery is especially beneficial demonstrating typically higher charge carrier mobility compared with that observed with an armchair periphery. Furthermore, zigzag carbon nanostructures exhibit strong absorptions in the visible region as a result of a decreased highest occupied molecular orbital–lowest unoccupied molecular orbital (HOMO–LUMO) gap [12].

2.1.1 Corrugations

The asymmetric distribution of carbon–carbon bond lengths resulting from the localized π electrons forces the graphene 2D lattice to become non-planar to minimize free energy, thus leading to the formation of ripples with heights up to 1 nm; see Fig. 2.4a. Qualitatively, the temporal and spatial modulation of the C–C bond lengths due to thermal vibrations and interatomic interactions induce carbon to occupy space in the third dimension [13]. Due to the enhanced asymmetry of the bond lengths near the edges or defects, the density of the ripples is higher in these regions [17, 18]. The amplitude of ripples is limited by the strain energy induced by its perpendicular displacement and increases with the size of the graphene sheet [19]. The distance between the ripples can be derived from [17]

$$L = 4\pi\xi\sqrt{\frac{2\pi}{3BT}}, \tag{2.4}$$

where ξ is the bending rigidity, B the two-dimensional bulk modulus and T the absolute temperature. The distance L between two ripples is therefore inversely proportional to the square root of the temperature and diverges to infinity at the temperature of 0 K.

The out-of-plane displacement ζ_r of the ripples can be well described by a sinusoidal function, $\zeta_r = A_r \sin(2\pi y/\lambda_r)$, where A_r and λ_r are the ripple amplitude and wavelength, respectively. Mechanistically, ripples in elastic thin films may be induced either by transverse compression in the y-direction or by longitudinal strain and/or shear in the x-direction [20]. Then, for a thin film of thickness t with clamped boundaries at $x = 0$ and $x = L$, in the presence of a longitudinal tensile strain [20]

$$\frac{A_r\lambda_r}{L} = \sqrt{\frac{8v}{3(1-v^2)}t}, \tag{2.5}$$

where v is the Poisson ratio ranging from 0.1 to 0.3 for single-layer graphene [21, 22]. For the in-plane shear strain [23],

$$\frac{A_r\lambda_r}{L} = \sqrt{\frac{8}{3(1+v)}t}. \tag{2.6}$$

Wrinkles and crumples are the other two types of corrugations in graphene sheets; see Fig. 2.4b and c. Unlike ripples having a modest aspect ratio of $\sim$1 and feature sizes of peaks and valleys below 10 nm, wrinkles exhibit a high aspect ratio of 10, and the width ranges from 1 to 10 nm, the height below 15 nm and the length above 100 nm [24]. Experimental results showed that there is considerable wrinkling in graphene and hence considerable out-of-plane deflection caused by the in-plane compression or shear. When wrinkling takes place, strain energy is stored within graphene nanosheets which is not sufficient to allow them to regain their shape. The wrinkles in the nanosheets are sundering apart at different locations while getting closer in other regions. As the nanosheets do not store sufficient elastic strain energy, wrinkling is an irreversible phenomenon but can be altered by external agents [25]. It was reported that controlled and organized microscale wrinkles could be produced by thermal manipulation, leveraging the negative thermal expansion coefficient of graphene [15]. Thermal expansion and reduction of graphite oxide have been widely used to synthesize crumpled graphene [26, 27].

The wrinkle wavelength λ_w and amplitude A_w for the free-edge thin film, such as suspended graphene on a trench can be calculated by [28]

$$\lambda_w^4 \approx \frac{4\pi^2 v L^2 t^2}{3(1 - v^2)\epsilon}, \tag{2.7}$$

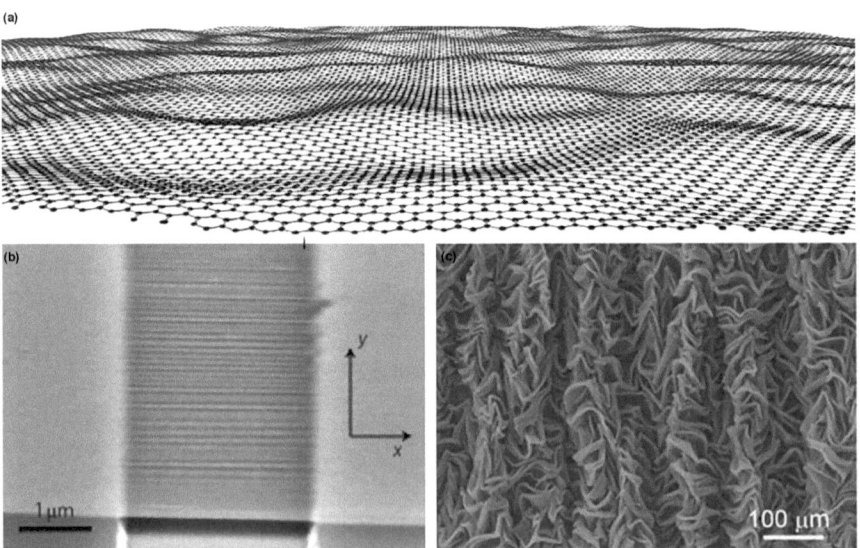

Fig. 2.4 **a** Rippled graphene, **b** wrinkled graphene and **c** crumpled graphene (reused with permissions from [13] Copyright ©2015 The Authors. Published by Elsevier Ltd., [14] Copyright ©2007, Springer Nature Limited, [15] Copyright ©2009, Springer Nature Limited, [16] Copyright ©2014, The Author(s))

$$A_w^4 \approx \frac{16\nu L^2 t^2 \epsilon}{3\pi^2(1-\nu^2)}, \qquad (2.8)$$

where ϵ is the edge contraction on the suspended graphene sheet.

In summary, the ripples on monolayer graphene can be formed due to the balance between the intrinsic and perpendicular restoring forces, the absence of other supporting (reinforcing) layers, and deformation-induced strains in graphene. The pattern of ripples depends on the temperature and size of the graphene sheet. Wrinkles are formed due to a uniaxial exterior force arising from the supporting layer of graphene. Further, crumple formation is a consequence of multidirectional forces applied to graphene. They depend on the original size of the sheet and the compression force on the graphene sheet.

2.1.2 Defects in Graphene

Theoretical studies have shown that introducing point defects, which involve non-hexagonal rings, into graphene layers can alter the mechanical, electronic, and magnetic properties of the sp^2 hybridized carbon network [6, 29, 30]. Under certain conditions, point defects can evolve into a one-dimensional metallic wire [31] or a two-dimensional semiconductor [32], which can be applied in graphene nanoelectronics.

The simplest defect in any material is the missing lattice atom that forms a single vacancy. The vacancy in graphene has been experimentally observed by transmission electron microscopy (TEM) [33, 34] and scanning tunneling microscopy (STM) [35]. The single vacancy experiences the Jahn–Teller distortion which saturates two of the three dangling bonds pointing them toward the missing atom; see Fig. 2.5. One of the bonds remains due to the geometry features.

The coalescence of two single vacancies or extraction of two neighboring atoms forms a double vacancy, as shown in Fig. 2.6. As seen in Fig. 2.6a, in a fully reconstructed double vacancy no dangling bond is present, while two pentagons and one octagon appear instead of four hexagons in perfect graphene. As a result, only minor perturbations appear in the bond lengths around the defect. Simulations indicate that

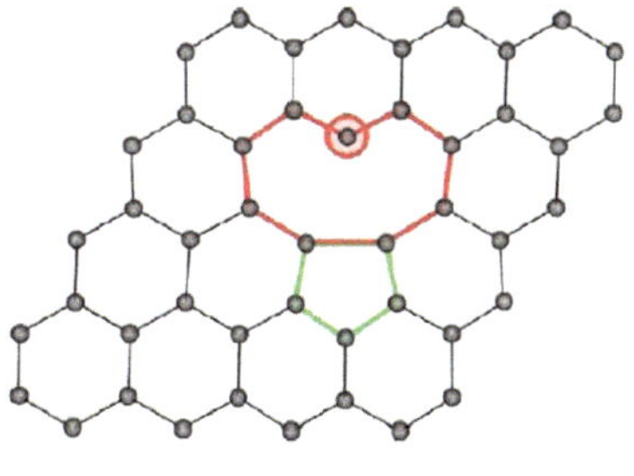

Fig. 2.5 Atomic structure of the single vacancy obtained from the density functional theory (DFT) calculations (reused with permissions from [29] Copyright ©2011, American Chemical Society)

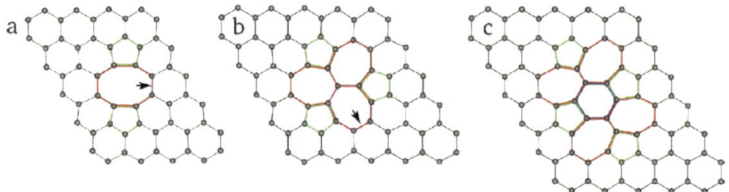

Fig. 2.6 Atomic structures of reconstructed double vacancy defects in graphene obtained from the DFT calculations (reused with permissions from [29] Copyright ©2011, American Chemical Society)

double vacancies are thermodynamically favored over single vacancies [36]. The activation energy for the migration of a double vacancy is about 7 eV [37], which is much higher than that for a single vacancy ($\approx$1.5 eV). This makes double vacancies in practice immobile up to very high temperatures.

The removal of more than two atoms leads to more complex multiple vacancies. As an important general remark, the vacancies with an even number of missing atoms and thus a complete saturation of dangling bonds are energetically favored over defect configurations with an odd number of missing atoms where an open bond remains [38].

Contrary to 3D crystals, interstitial atoms do not exist in graphene since the placement of an atom to an in-plane position (e.g., in the hexagon center) requires an exceedingly high energy. It is inferred from general considerations that bonding in the bridge configuration, on top of a C–C bond, is energetically favored [29]. Some degree of sp^3 hybridization can appear locally when a carbon atom interacts with a perfect graphene layer so that two new covalent bonds can be formed between the adatom and the underlying atoms in the graphene plane. The binding energy of the carbon adatom is of the order of 1.5–2 eV [39, 40].

Topological structural defects of the honeycomb lattice (substitution of a hexagonal ring by other polygons), such as disclinations, dislocations, pentagons, heptagons, Stone–Wales defects (special dislocation dipoles formed by a combination of two pentagon-heptagon pairs), induce long-range deformations, which modify both the phonon and electron scattering processes [6, 29].

A disclination is equivalent to the deletion or inclusion of a wedge in the lattice. The simplest one in the honeycomb lattice is the absence of a 60° wedge. The resulting edges can be glued so that all sites remain threefold-coordinated. The honeycomb lattice is recovered everywhere, except at the wedge's apex, where a fivefold ring, a pentagon, is formed.

The configuration of disclinations resulting from the addition or removal of semi-infinite wedges is exemplified in Fig. 2.7. Thus, the positive wedge angle ($s = 60°$) allows a pentagon to be embedded into the honeycomb lattice of graphene, while the negative wedge angle ($s = -60°$) creates a heptagon embedded in the graphene lattice. The isolated non-hexagonal rings in graphene inevitably result in non-planar structures.

positive disclination   ideal graphene   negative disclination
$s = 60°$ $s = -60°$

Fig. 2.7 Positive ($s = 60°$) and negative ($s = -60°$) disclinations in graphene are produced by either removing or adding a 60° wedge (shaded area) of material without changing the coordination of carbon atoms (reused with permissions from [41] Copyright ©2010 American Physical Society)

Dislocations are important to account for the mechanical and electrical behavior of materials. In graphene, mechanical characteristics can have striking consequences for its electronic properties [42, 43]. One-dimensional defect structures embedded in graphene have been treated computationally [41, 43, 44]. A one-dimensional defect has also been found experimentally in graphene [44–48], resembling dislocations in a conventional 3D crystal. Screw dislocations require a three-dimensional strain field that does not exist in graphene. Meanwhile, an equivalent of an edge dislocation can be imagined in graphene, but only in its projection onto a plane because no dislocation line normal to the layer exists. So far, however, no one-dimensional defect in graphene has been convincingly observed in experiments.

Line defects in graphene frequently separate domains of different crystallographic orientations. Particular examples could be obtained in graphene growth on metal surfaces [31], for example, due to lattice mismatch in graphene grown on a Ni surface (see Fig. 2.8). This defect consists of an alternating line of pairs of pentagons separated by octagons, as shown in Fig. 2.8a.

The delocalized π electron on the surface of graphene sheets can readily form strong $\pi - \pi$ stacking interactions with aromatic compounds. This excellent chemical reactivity of graphene together with its relatively large specific surface area makes it a promising adsorbent [49–52]. Several important points are in order in this context.

First, a vast literature has developed around the intrinsic properties of the graphene surface dramatically affected by contamination of surface-active sites [53–61]. Different polar groups such as H, F, CH_3, or OH bind covalently to carbon atoms, transforming the trigonal sp^2 orbital to the tetrahedral sp^3 orbital [62].

Second, the trace metallic and metalloid impurities [63], acidic residues [64], hydrocarbons [53] and even silicon [65] are ubiquitous on graphene because top-down procedures of its synthesis (which consists of the successive oxidation/ exfoliation/reduction steps and requires graphite as starting material) are prone to contamination originating from the starting synthetic graphite with of 98.0–99.9% wt. [53].

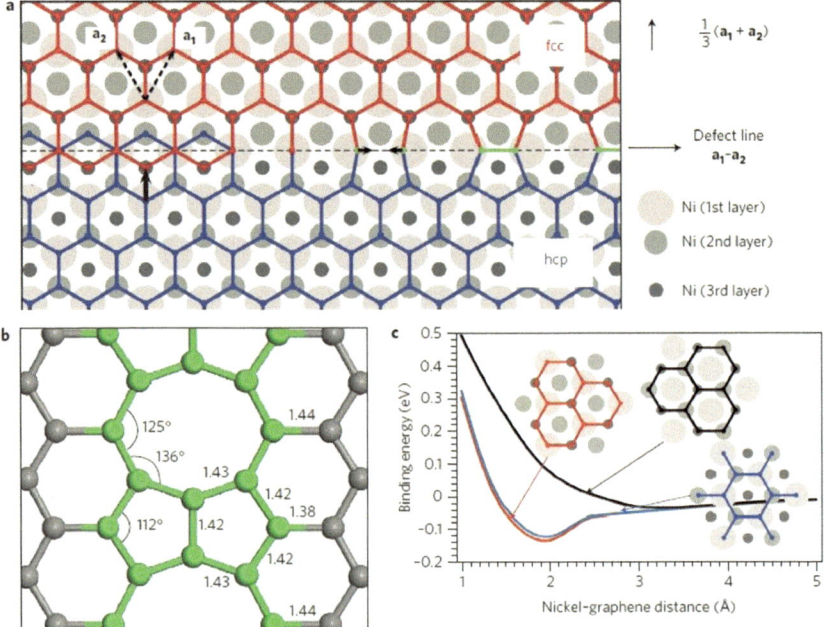

Fig. 2.8 Schematics of a one-dimensional defect in graphene grown on a Ni surface. **a** Two graphene half-lattices, with unit-cell vectors $\mathbf{a}_1$ and $\mathbf{a}_2$, are translated by a fractional unit-cell vector 1/3 $(\mathbf{a}_1 + \mathbf{a}_2)$, indicated by the vertical vector (solid arrow). The two half-lattices can be joined along the $\mathbf{a}_1 - \mathbf{a}_2$ direction, indicated by the horizontal vector, without any unsaturated dangling bonds, by restructuring the graphene lattice. The domain boundary can be constructed as shown, by joining two carbon atoms, indicated by the two arrows, along the domain boundary line. This reconstructed domain boundary forms a periodic structure consisting of octagonal and pentagonal carbon rings. The underlying Ni(111) structure illustrates how the extended defect is formed by anchoring two graphene sheets to a Ni(111) substrate at slightly different adsorption sites. If one graphene domain has every second carbon atom located over a fcc-hollow site (red) and the other domain over a hcp-hollow site (blue), then the two domains are translated by $1/3(\mathbf{a}_1 + \mathbf{a}_2)$ relative to one another. **b** Relaxed geometry of the defect structure obtained by using the DFT and showing the bond lengths (in Å) and bond angles. **c** The calculated adsorption energies for these two domains are very similar, but both are lower in energy than a third possible adsorption configuration with all carbon atoms on hollow sites (reused with permissions from [31] Copyright ©2010, Springer Nature Limited)

Third, the graphene surface can be conjugated with a variety of ionically and covalently bonded impurities [67]. Chemical bonding of impurities like hydrogen or fluorine on a graphene sheet may generate a local distortion of the hexagonal lattice and lead to spin–orbit coupling [66]. The covalent impurities with one chemically active electron cause universal midgap states as the carbon atom next to the impurity is effectively decoupled from the graphene π bands. The electronic structure of graphene suppresses the migration of these impurities and makes the universal midgap very stable. This effect is strongest for neutral covalently bonded impurities.

The ionically bonded impurities have migration barriers of typically less than 0.1 eV, which is about an order of magnitude less than their typical binding energies [67].

2.2 Multilayer Graphene and Graphene Derivatives

The term graphene is usually referred to as isolated monolayer graphene [68] and can also be extended to bilayer graphene, as both of them are semimetals with no overlap between the valence and conduction bands [69]. The electronic structure of few-layer graphene (FLG) with layer numbers from 2 to about 5 is more complex because of the appearance of charge carriers. It has been shown that the electronic structure of graphene rapidly evolves with the number of layers, approaching the 3D limit of graphite at 10 layers [70]. When stacked with a greater number of layers (typically up to 10), multilayer graphene (MLG) is obtained. A general scheme of monolayer, few-layer, multilayer graphene together with bulk graphite is sketched in Fig. 2.9.

The properties of graphene depend on the number of stacked layers. Most generally, monolayer graphene has better mechanical, thermal, and electrical properties than few-layer or multilayer graphene [71]. However, the synthesis of monolayer graphene is complicated and expensive [72, 73]. Therefore, synthesizing few-layer graphene or multilayer graphene is also a subject of interest [74].

The stacking arrangements in bilayer graphenes can be either AA or AB. For AA stacking, each carbon atom in the top layer is directly aligned on the top of the appropriate C-atom in the bottom layer. In contrast, in bilayer graphene with AB (or Bernal) stacking, a set of atoms in the top layer sits over the empty centers of hexagons in the bottom layer.

In multilayer graphene, the stacking orders can become more complicated. Experimentally and computationally, there are two stable stacking arrangements, Bernal (ABA order) and rhombohedral (ABC sequence), as shown in Fig. 2.10 for trilayer graphene.

In thermodynamic terms, Bernal structures appear to be a little bit more stable. In both structural configurations, the stacking of graphene layers is attributed to the

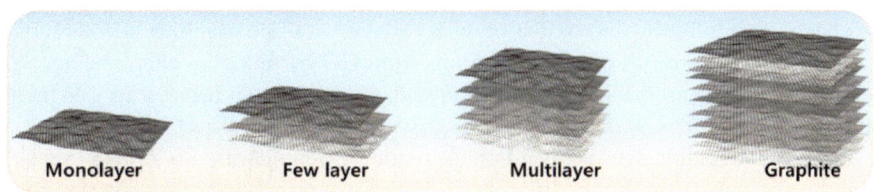

Monolayer   Few layer   Multilayer   Graphite

Fig. 2.9 Schematic sketch of the monolayer, few-layer, and multilayer graphene in comparison with graphite (reused with permissions from [71] Copyright ©2022 by the authors. Licensee MDPI, Basel, Switzerland)

Fig. 2.10 Bernal **a** and rhombohedral **b** stacking arrangements in trilayer graphene (reused with permissions from [7] Copyright ©2018 The Author(s). Published by National Institute for Materials Science in partnership with Taylor & Francis Group)

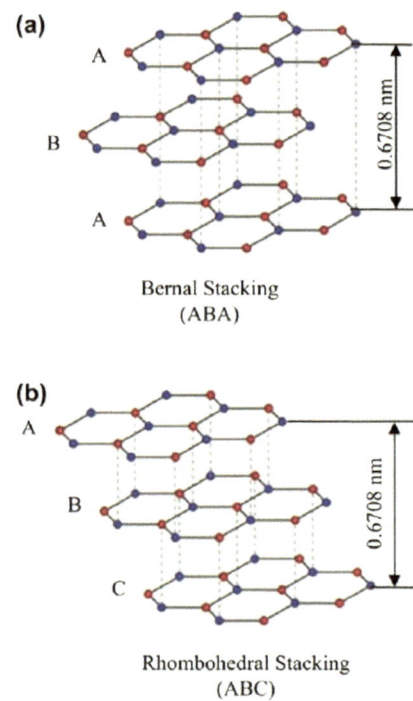

Bernal Stacking
(ABA)

Rhombohedral Stacking
(ABC)

weak interaction between π bonds in the adjacent basal planes. Both ABA and ABC sequences appear to have the same distance of 0.3354 nm between carbon sheets.

2.2.1 Graphene Oxide

Starting from pristine raw graphene, different derivatives have been synthesized, among which the most versatile are graphene oxide (GO), reduced graphene oxide (rGO), graphene nanoribbons, graphene quantum dots, or metal oxide doped graphene with wide-ranging structures and characteristics [75]. Graphene oxide results from the chemical exfoliation and oxidation of layered crystalline graphite, while rGO is derived from the removal of oxygen groups from GO by reducing chemical agents, such as hydrazine, ascorbic acid, and many others, or through thermal or UV treatment of GO [76].

The most adaptable derivate is graphene oxide. Among numerous strategies used, the reduction of GO may be considered one of the most promising routes. Meanwhile, the reduction mechanism remains ambiguous and the detailed structure of GO is still a subject of debate [77–81]. The solubility of GO in water and other solvents allows it to be uniformly deposited onto different substrates in thin films or networks, making it potentially useful for macroelectronics [78].

The structure of GO is often simplistically assumed to be a graphene sheet covalently bonded to oxygen-containing functional groups. These functional groups include epoxide (=C–O–C=) and hydroxyl (OH) groups on the basal plane and carboxylic acid (–(C=O)–OH), carbonyl (–C=O), and other ketone groups (R_1–C(=O)–R_2) groups mostly on the edges [83–85]. Oxirane and alcohol groups have also been identified on the GO surface [85]. The contents of these functional groups can vary significantly, depending on the starting material and oxidative conditions. The level of oxidation (C/O ratio) plays a major role in determining the affinity of graphene derivates with organic solvents and polymer matrices in nanocomposites [86]. The relative percentages of the main functional groups in GO and rGO have been determined from the X-ray photoelectron spectroscopy (XPS) measurements. The results are summarized in Table 2.1.

As there are many oxygen-containing functional groups on the graphene oxide sheet layer compared with graphene, the structure of GO is more complicated. The final structure of GO obtained depends on the chemical oxidation process used. The essential structural models employed are the Lerf–Klinowski (L-K) model [82] updated by Gao et al. [77] and the Szabó-Dékány model [83]. The L-K model suggests that the hydroxyl and epoxy groups are randomly distributed on the single-layer of GO, while the carboxyl and carbonyl groups are attached at the edge of the layer. Graphene oxide has an unoxidized benzene ring region and an oxidized aliphatic six-membered ring region, and the relative size of these two regions depends on the degree of oxidation and random distribution on the GO layer.

In the Dékány model, the tertiary OH groups and 1,3-ethers lie above and below the cyclohexane sheets, while cyclic ketones and quinones can be formed on the hexagon ribbons where the C–C bonds are cleaved. It is also possible to include phenolic groups into the bulk of the layers, by which the planar acidity can be easily explained without assuming unstable enols and C–H bonds in GO.

Other models have also been proposed, such as Hofmann–Holst model [87], Ruess model [88], Scholz–Boehm model [89], Nakajima–Matsuo model [90], dynamic structure model [91], and binary structure model [92]. Some of the models are summarized in Fig. 2.11.

Most recently, Tao et al. [81] proposed a new GO model by integrating potentiometric titrations and ab initio calculations to mimic the oxidation process (Fig. 2.12).

Table 2.1 Summary of the relative percentages of the carbon and carbon–oxygen bonds in GO and rGO [85]

Binding energy (eV)	284.5	286.0–286.7	287.6–287.8	289.0–289.4
Bond assignation	C–C	C–O	–C=O	O–C=O
GO	48.6	36.8	9.5	5.2
rGO	75.5	16.7	3.5	4.3

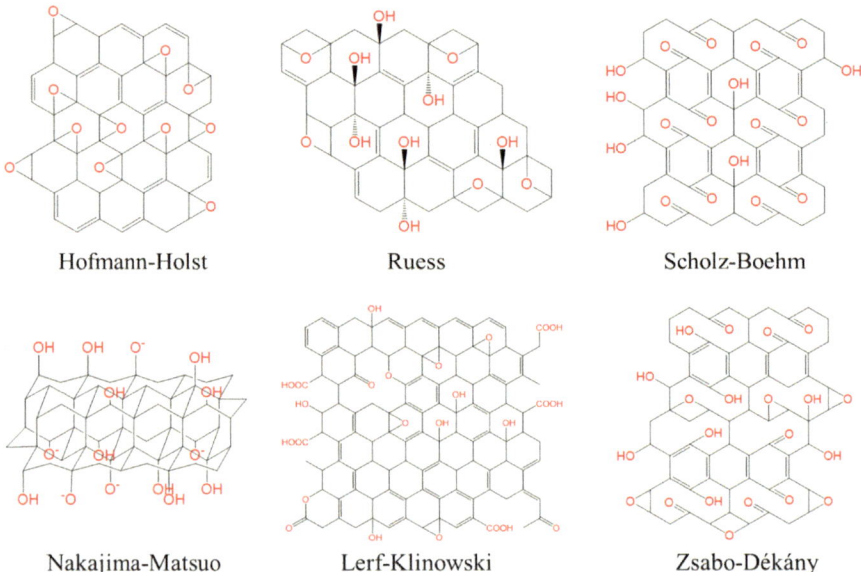

Fig. 2.11 Six proposed structural models of graphene oxide (reused with permissions from [81] Copyright ©2023 The Authors. Published by American Chemical Society)

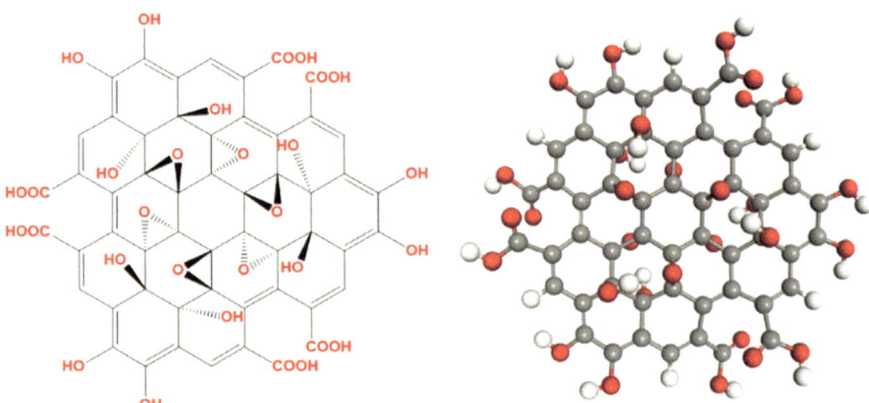

Fig. 2.12 Graphene oxide modeled by assembling a series of GO fragments with a central carbon ring in the highest oxide state. The gray, white, and red spheres represent the carbon, hydrogen, and oxygen atoms, respectively (reused with permissions from [81] Copyright ©2023 The Authors. Published by American Chemical Society)

Functional groups at the edges are mainly double-interactive carboxyls and double-adjacent phenolic hydroxyls, while groups on both sides of the plane are mainly collocated epoxies and hydroxyls with a meta-positional hydrogen bond interaction.

Ideally, a graphene sheet consists of trigonally bonded sp^2 carbon atoms and is perfectly flat provided the roughness of the graphene surfaces may simply reflect the corrugations of the supporting substrate [93]. Moreover, the existing intrinsic ripples in graphene can be strongly suppressed by interfacial Van der Waals interactions when the sheet is supported on an appropriate atomically flat substrate. In contrast, the decorated GO sheets consist partly of tetrahedrally bonded sp^3 C atoms, which are displaced slightly above or below the graphene plane [26]. Due to the structure deformation and the presence of covalently bonded functional groups, GO sheets are atomically rough [78, 94]. Experiments indicate that the degree of oxidation fluctuates at the nanometer scale, suggesting the presence of sp^2 and sp^3 carbon clusters of a few nanometers with highly defective regions [78, 94, 95]. A surprisingly perfect graphene-like honeycomb lattice is, nevertheless, seen in GO even though the carbon atoms attached to functional groups are slightly displaced. Consequently, the overall size of the unit cell in GO remains similar to that of graphene [96]. Graphene oxide can therefore be described as a random distribution of oxidized areas with oxygen-containing functional groups, combined with non-oxidized regions where most of the carbon atoms preserve sp^2 hybridization.

Irrespective of the exact structure, the obvious difference between graphene and GO elucidated in extensive research is the addition of oxygen atoms bound to carbon as shown in Figs. 2.11 and 2.12. Consequently, graphene is hydrophobic, whereas graphene oxide is hydrophilic and easily dispersed in water. As mentioned above, graphene oxide can be reduced to graphene by a reducing agent. However, the resulting graphene is not suitable for electronic applications and mechanical reinforcement of polymers because of the structural defects that occur in the synthesized GO. Meanwhile, this route is preferable for the large-scale modification of the surface properties of graphene materials by functionalization [97].

Having hydrophilic functional groups (–OH, epoxide, –COOH), which promotes the intercalation of water molecules into the gallery, the graphene sheets can be easily detached from each other by sonication, thus producing highly dispersible GO sheets in aqueous medium [98]. However, a particular challenge to work with resulting GO sheets is that they have a high tendency to form agglomerates due to the Van der Waals forces and strong $\pi - \pi$ stacking [99].

2.2.2 Reduced Graphene

As described above, the attachment of oxygen-containing functional groups to graphene sheets during chemical oxidation changes the hybridization of the carbon atoms from sp^2 to sp^3. The electron and phonon transport in graphene is defined by the long-range conjugated network of the graphitic lattice [100, 101]. Oxygen-containing functional groups break the conjugated bridges and localize π-electrons, resulting in decreased electron and phonon mobility and free carrier concentration. Although there are conjugated regions in GO, long-range conductivity is blocked by the absence of percolating pathways between sp^2 carbon clusters.

Consequently, grown GO is typically insulating with a sheet resistance of greater than 10^{12} Ω/square [102, 103].

The electrical conductivity of GO is typically enhanced through the removal of the oxidized moieties in the sheets by chemical reduction yielding rGO [104]. The reduction of GO is not aimed only at removing the oxygen-containing groups and other lattice defects thus partly restoring the structure and properties of graphene but also at recovering the conjugated network of the graphitic lattice. These structure changes result in the recovery of electrical conductivity and other properties of graphene. The existence of countless electrically active sites in rGO in turn makes it a better candidate for sensing applications and imparts high structural resemblance with graphene [105]. However, the reduced GO inevitably contains lattice defects that degrade its electrical properties compared to pG sheets. Thus, chemically reduced films of a few layers GO with sheet showed a resistance of about 4 MΩ/square [106], while 800 kΩ/square were obtained in 25 nm thick nanocomposite films of silica/GO [107]. In comparison, a few-layer graphene sheet with a thickness of less than 3 nm has a sheet resistance of about 400 Ω/square at room temperature [108], while a four-layer film ($\sim$2 nm) gives about 30 Ω/square [109]. Since the main purpose of GO reduction is to restore the high conductivity of graphene, the electrical conductivity of rGO is frequently used to determine the effectiveness of the reduction method.

The reduced GO sheets are usually considered as one kind of chemically derived graphene. Some other names are also used for reduced graphene, such as functionalized graphene, chemically modified graphene, and chemically converted graphene [110]. The reduction of GO makes the material attractive for numerous applications. The possibility for synthesis of large-area rGO sheets with good electrical conductivity provides the way for the electrical property-related applications of large-area GO sheets.

There is an obvious advantage of a better dispersion of GO when it is used as a filler in the production of polymer nanocomposites [111]. Indeed, in numerous applications such as optical electronics, energy conversion and storage materials, catalysis, and sensors, graphene derivatives have to be blended with other components to form functional composites (see Sect. 3). rGO-based composite materials are of special importance because of their intriguing properties originating from the synergistic effects of their components over the intrinsic properties of each partial component. It is obvious from the above discussion that pristine graphene is usually insoluble, intractable, and decomposes before melting. Therefore, it cannot be simply blended with other components by conventional processing techniques into uniform composites. Instead, when the GO sheets are reduced into reduced graphene oxide by careful removal of the excessive oxygen groups, the resulting rGO can effectively be used as fillers in the graphene-based polymer nanocomposites.

There are many different strategies for the reduction of GO to rGO which can be divided into three classes: chemical reduction, thermal reduction, and electrochemical reduction [105, 112–116]. The chemical reduction method is used to obtain large rGO surface areas and superb electrical conductivity [113, 117]. It can employ some reducing agents like NaBH$_4$, hydrazine, NaOH, Na$_2$CO$_3$ and L-ascorbic acid.

The production of rGO through a thermal reduction process yields bulk amounts of rGO in a short reduction time [113, 117, 118]. However, the heating can damage the graphene platelets and also incorporate imperfections and vacancies affecting the mechanical strength and electrical conductivity of rGO. In the electrochemical approach, the ITO or glass substrates are covered with a layer of GO, and a current is passed through the layer. The resulting rGO has a high carbon-to-oxygen ratio and demonstrates excellent conductivity [119]. It is important to recognize that the electrochemical reduction provides growth of high-quality materials so that the quality of the electrochemically produced rGO is comparable with that of pure graphene [120].

Some other approaches were also proposed such as a vapor treatment, toughening, laser, and microwave reduction. A hydrothermal method was also applied to remove the oxygen groups [121]. An rGO was also synthesized via ultrasonic exfoliation and chemical reduction with hydrazine hydrate [104]. Commercial methods to produce nanoplates of reduced graphene oxides were also developed [113, 117, 122, 123]. The transformation of GO into rGO is schematically illustrated in Fig. 2.13.

2.2.3 Graphene Nanoplatelets

Graphene nanoplatelets (GNPs) are a new form of carbon species formed from graphite under certain conditions. GNPs are microscopic particles that, in at least one dimension, measure less than 100 nm and bridge atomic structures and bulk materials. These nanoparticles are mostly found in the range of 1–15 mm in thickness and up to 100 μm in lateral size. The unique size and morphology of these nanoscale stacks of platelet-shaped graphene sheets give them the ability to easily disperse into other materials, thus creating higher value composite materials that are enhanced by the exceptional thermal, mechanical, and electrical properties of graphene. There is still a lack of large-scale manufacturing techniques for pure graphene because of the low fabrication rates and high sales costs. GNPs combine large-scale production and low costs with remarkable physical properties [124].

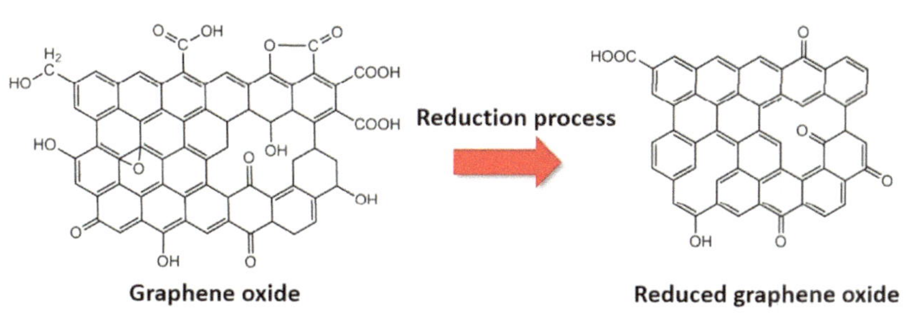

Graphene oxide   Reduced graphene oxide

Fig. 2.13 Illustration of the reduction process from graphene oxide to reduced graphene oxide (reused with permissions from [118] Copyright ©2016 Elsevier Ltd.)

Graphene-based materials are classified according to their thickness, lateral size, and carbon-to-oxygen atomic ratio [68, 125]. Commercially available GNPs are a mixture of single-layer, few-layer (2–10 layers), and nanostructured graphite. Therefore, GNP thickness can vary from 0.34 to 100 nm within the same production batch [126, 127].

The GNP nanoflake powder is usually obtained following the liquid-phase exfoliation procedure without further centrifugation steps. Other manufacturing methods include ball-milling [128], the exposure of acid-intercalated graphite to microwave radiation [128], shear-exfoliation, and the wet-jet milling [129]. These manufacturing techniques produce a large variety of powders in terms of thickness, the lateral size of the flakes, aspect ratio, and defect concentrations [129]. GnPs are typically composed of single and few-layer graphene mixed with thicker graphite; see Fig. 2.14.

The synthesis of GNPs offers the development of graphene nanocomposites with enhanced barrier properties [130]. GNPs have a percolation threshold for conductivity of 1.9 wt.% in a thermoplastic matrix. They can also enhance the mechanical characteristics, including stiffness and tensile strength of different composites due to the strong interfacial interaction of nanoplatelets with the matrices.

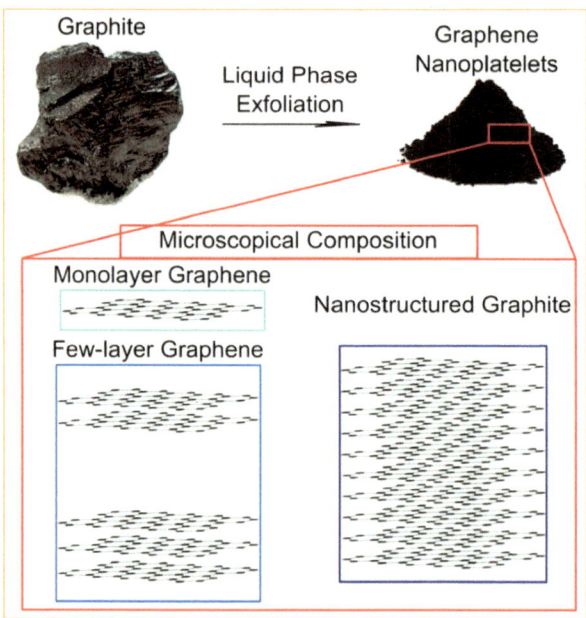

Fig. 2.14 Schematic of the fabrication of GNPs starting from natural graphite. The typical black powder obtained after liquid-phase exfoliation and solvent evaporation is constituted by a mixture of single and few-layer graphene and nanostructured graphite (reused with permissions from [124] Copyright ©2018 by the authors. Licensee MDPI, Basel, Switzerland)

2.2.4 Graphene Quantum Dots

Another example of the graphene derivatives known as graphene quantum dot (GQD) is considered to be a zero-dimensional (0D) carbon-based material, defined as small pieces or fragments of graphene with lateral dimensions smaller than 100 nm in single or a few layers [131–133]. GQDs are of immense interest for their potential substitution of semiconducting quantum dots in various applications ranging from photovoltaic to drug delivery and biosensing [134]. One of the essential features of GQDs is the presence of graphene crystal lattice [135, 136]. This is in stark contrast to more versatile carbon quantum dots, or simply carbon dots (CDs), which are quasi-spherical particles less than 10 nm in size with less hybridized sp^2 carbons [133, 137]. GQDs are anisotropic and have a layered structure, While CDs have a nanoparticle structure and a colloidal character [136, 138].

The structure of GQDs depends upon the fabrication conditions which have been widely discussed in the literature [136]. Anyway, by the above definition, GQDs are composed of a graphene crystal lattice with carbon atoms arranged in honeycomb rings of six atoms. Each atom is covalently bonded to three other carbon atoms, which gives the sp^2 hybridized characteristic and results in delocalized electrons in the π orbitals perpendicular to the plane of the sheet. Such features are responsible for their outstanding properties like low toxicity, enhanced photoluminescence, chemical stability, and strong quantum confinement [136, 139].

There are two types of edges in GQDs, armchair and zigzag [133, 140], as shown in Fig. 2.15. When a graphene sheet is cut along an armchair line, triple carbon bonds are formed at the edges, as shown in Fig. 2.15a. This yields a carbyne-like structure, which may have a non-polar characteristic due to carbon triple bonds at the edges. Along a zigzag line, two unshared valence electrons are made at each edge carbon atom, as shown in Fig. 2.15b yielding a carbene-like structure, which may have a polar characteristic due to lone-pair electrons. The edge types affect the shape of the GQDs and, consequently, their electronic and optical properties. A 120° corner is made when two armchair lines or two zigzag lines are encountered, while a 90° corner is made when armchair and zigzag lines are encountered. Therefore, in tailoring a graphene sheet, hexagonal GQDs are made when the same type of cutting lines are encountered at all corners, while rectangular GQDs are made when different types of cutting lines are encountered at all corners; see Fig. 2.15c. However, circular or oval shapes can also be realized when the corners cannot be well developed [140].

GQD fabrication techniques can be divided into (i) top-down approaches when GQDs are obtained by fragmentation or exfoliation of graphene sheets using chemical, physical, and electrochemical methods, and (ii) bottom-up approaches when small organic molecules are used as precursors to obtain GQDs. One example of the bottom-up approach is shown in Fig. 2.16. The pyrolysis of precursors such as citric acid produced GQDs of about 15 nm in diameter and 0.5–2.0 nm in thickness by modifying the carbonization conditions and dispersing the pyrolyzed products in alkaline media [131]. Complete carbonization achieved by prolonged heating of

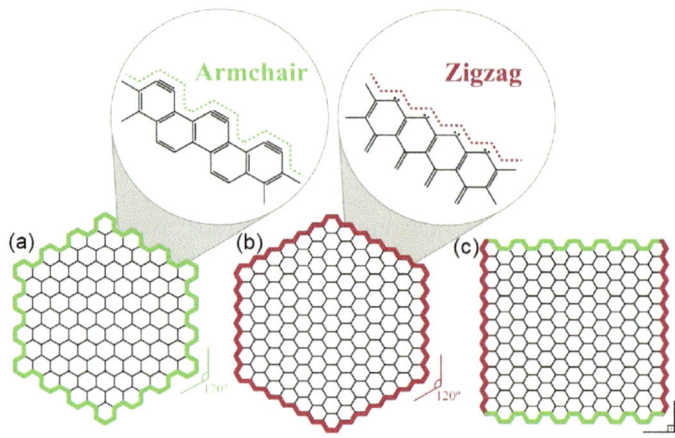

Fig. 2.15 The shapes of GQDs with the armchair **a**, zigzag **b**, and hybrid armchair–zigzag **c** edges (reused with permissions from [133] Copyright ©2020 the Royal Society of Chemistry)

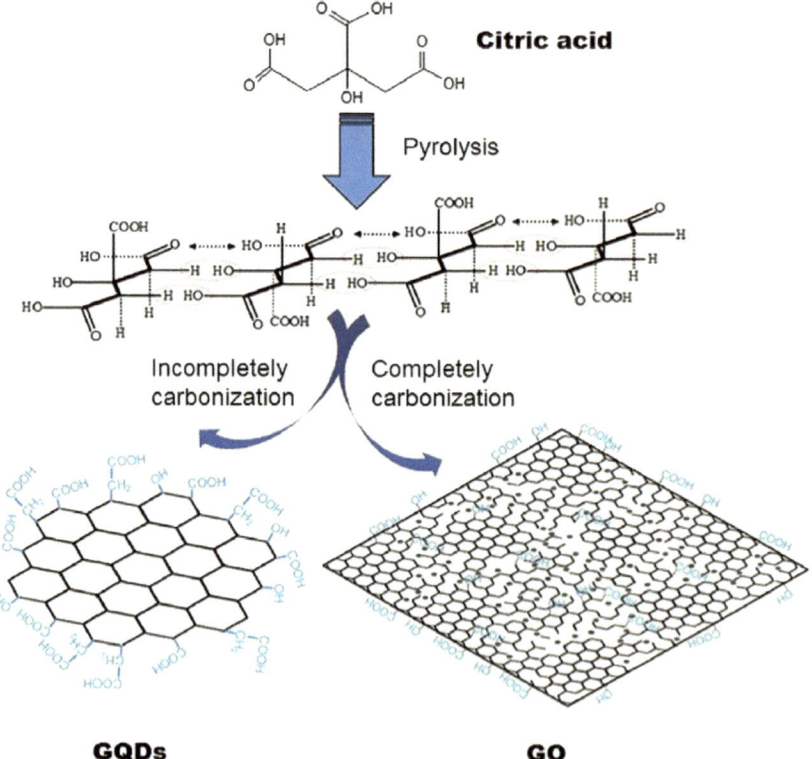

Fig. 2.16 Schematic of the fabrication of GQDs and GO by pyrolysis technique using citric acid as a raw material (reused with permissions from [131] Copyright ©2012 Elsevier Ltd.)

the precursors produced GO nanostructures, about 100 nm in width and 1 nm in thickness.

2.2.5 Surface Functionalization

Despite the great interest in applications, graphene itself has zero band gap and chemical inertness. This weakens the competitiveness of graphene in the field of semiconductors and sensors. Thus, to overcome this problem, there has been a huge interest in the functionalization of graphene including reactions of graphene and its derivatives with organic and inorganic molecules, chemical modification of the large graphene surface, and the general description of various covalent and noncovalent interactions with graphene [141]. Band gap opening of graphene by doping, intercalation, and striping is useful for designing nanoelectronic devices. Here, the functionalization of graphene and its derivatives are only briefly described. Instead, we will be focusing in more depth on the opportunity of their functionalization with polymers.

Combining the robustness and flexibility of membrane, graphene provides essentially infinite possibilities for the modification and functionalization of its carbon backbone [142]. Meanwhile, the chemical inertness of pristine graphene makes it difficult to modify pG chemically. Limited reported methods involve reactive species such as nitrenes and free radicals released from diazonium ions, perfluorinated alkyl iodides, 2,2,6,6-tetramethylpiperidinyloxy (TEMPO), benzoyl peroxide, or perfluorophenylazides (PFPA) [143]. The addition of radicals to graphene, on the other hand, converts sp^2 carbons to sp^3 thus yielding a single covalent bond.

Free radical additions are among the most common reactions, and these radicals can be generated from diazonium salts and benzoyl peroxide. Electron transfer from graphene to aryl diazonium ion or photoactivation of benzoyl peroxide yields aryl radicals that subsequently add to graphene to form covalent adducts. Nitrenes, electron-deficient species generated by thermal or photochemical activation of organic azides, can functionalize graphene very efficiently. Carbenes are used less frequently than nitrenes but undergo CH insertion and C=C cycloaddition reactions with graphene. In addition, arynes can serve as a dienophile in a Diels–Alder type reaction with graphene [144].

The chemical functionalization of graphene by introducing atoms and atomic groups has been given increasing attention for its promising capabilities to modify physical and chemical properties [141, 145]. For example, due to the functional groups of the graphene, a homogeneous blend of graphene–polymer composite was designed for an organic photovoltaic device [146]. Functionalized graphene sheets were used as amperometric glucose biosensors [147]. These graphene sheets were prepared by exfoliating graphitic oxide with a decoration with crystalline (Pt-Au)/Au metal nanoparticles using a chemical reduction method. The erosion of hydrogen-functionalized graphene produces hydrocarbon molecules, such as CH_x and C_2H_x, which affect the plasma surface interaction [148]. GO-based materials have demonstrated extraordinary sensitivity to water molecules due to their hydrophilicity, which can be further improved via additional functionalization [149]. Previous studies suggest that the interaction between GO and water molecules results in the formation of a hydrogen bond network and modifies the interlayer structure of GO laminates. This interaction can be used for environmental applications such as moisture detection and atmospheric water harvesting. When a surface functionalized graphene separator was used in the Li-metal batteries, it exhibited superior electrolyte wettability compared to the graphene separator, contributing to the improved ionic conductivity and homogeneous Li-ion flux [150].

Many polymerization methods, e.g., atom transfer radical polymerization (ATRP), reversible addition-fragmentation chain transfer polymerization (RAFT), nitroxide mediated radical polymerization (NMRP), anionic polymerization, and ring-opening polymerization (ROP) techniques have been used to functionalize graphene [151].

Selected reaction types for the functionalization are illustrated in Fig. 2.17. They include hydrogenation, the addition of phenyl radicals $-C_6H_5$, diazonium species $R\text{-}N^+ \equiv NX^-$, where R can be any organic group like alkyl or aryl and X an inorganic or organic anion such as halide, 1,2,3-triazoles $C_2H_3N_3$ with a five-membered ring of two C atoms and three N atoms, azomethine ylides $R_1R_2C^- - N^+R_3 = CR_4R_5$, fluorinated phenylnitrene species $C_7N_4F_4OH$, aryne (highly reactive species derived from an aromatic ring by removal of two substituents), carbenes $R_1-:C-R_2$, and tetracyanoethylene $C_2(CN)_4$.

Frequently, either functionalized materials are not single-layered or few-layer graphene is functionalized which produces covalently functionalized few-layer graphene that can be isolated in stacks; see Fig. 2.18a. To overcome this obstacle for the functionalization of graphene, the activation of graphite before exfoliation provides an opportunity to address single layers to synthesize functionalized G1 graphene, even if stacks of functionalized G1 graphene (G_1-R_n) are isolated; see Fig. 2.18b.

The chemical functionalization of graphene was performed by using various reactants such as hydrogen, oxygen, or halogens, which produced partially functionalized graphene [152]. It should be mentioned that the complete hydrogenation of graphene leading to fully sp^3-hybridized C atoms has not yet been experimentally realized. The highest degree of functionalization, which approaches the 1:1 stoichiometry, was achieved by the reaction of graphene with xenon difluoride to form fluorinated graphane. The compound called graphane is a fully saturated hydrocarbon derived from a single graphene sheet with the formula CH [153]. All of the carbon atoms are in sp^3 hybridization forming a hexagonal network and the hydrogen atoms are bonded to carbon on both sides of the plane in an alternating manner (Fig. 2.19).

Particularly elegant is that changing the hybridization of carbon atoms from sp^2 into sp^3 removes the conducting p-bands and opens an energy gap. The change in hybridization from sp^2 into sp^3 generally results in longer C–C bonds, which is the effect opposing to the lattice shrinkage by atomic-scale buckling [154]. The charge carrier mobility of graphane is three orders of magnitude smaller than that of graphene; as a consequence, graphane behaves as an insulator. The locally flat graphene structure becomes buckled in graphane, the C–C–C angle decreases from

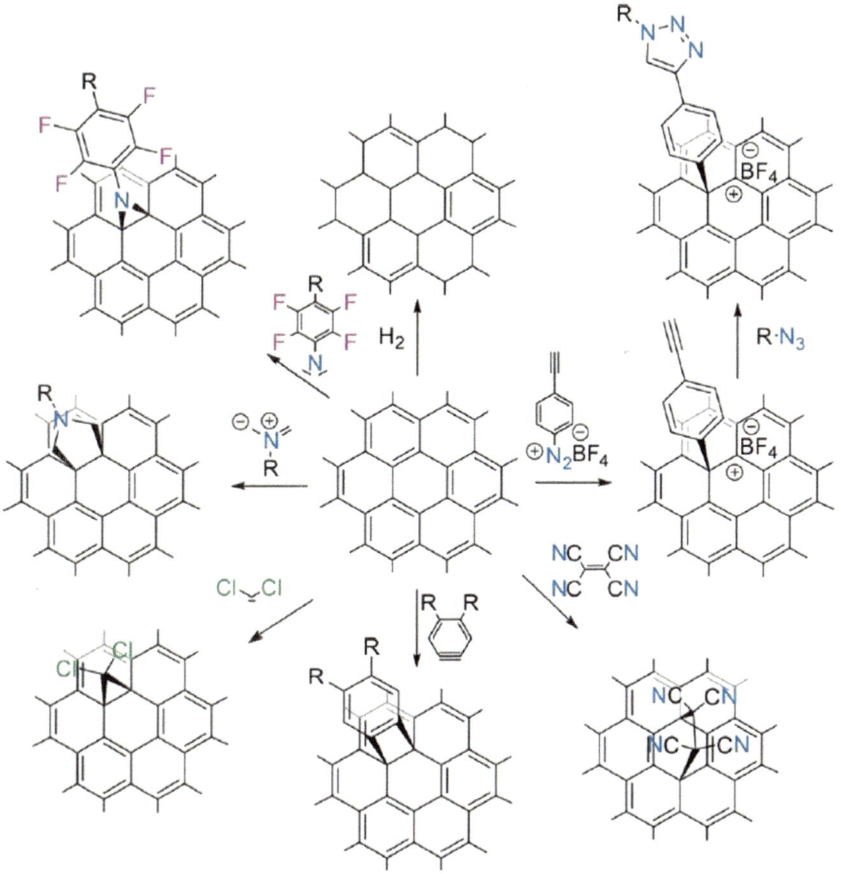

Fig. 2.17 Several typical reactions for the functionalization of graphene and FLG (reused with permissions from [152] Copyright ©2014 Wiley-VCH Verlag GmbH & Co. KGaA, Weinheim)

120° (in graphene) to 109.5° and the C–C bond increases from 1.42 Å (in graphene) to 1.52–1.56 Å [141]. Graphane is predicted to be stable with binding energy comparable to other hydrocarbons such as benzene, cyclohexane, and polyethylene.

Apart from free radicals, dienophiles also react with sp^2 carbons of graphene. Azomethine ylide, which reacts through a 1,3-dipolar cycloaddition, is one of the most common dienophiles successfully applied in the functionalization of graphene. Georgakilas et al. [155] showed that these graphene sheets could be substituted with pyrrolidine rings via a 1,3-dipolar cycloaddition of azomethine ylide. Zhao et al. [156] reported that the azomethine ylide-functionalized graphene (G-OH) could serve as a versatile platform for post-syntheses of various graphene-derived materials including epoxide-functionalized graphene (G-EP); see Fig. 2.20.

The 2-(3,4-dihydroxyphenyl) pyrrolidine (DHPP) grafted graphene, hereafter named as G-OH, is demonstrated to be just such a graphene derivative. As

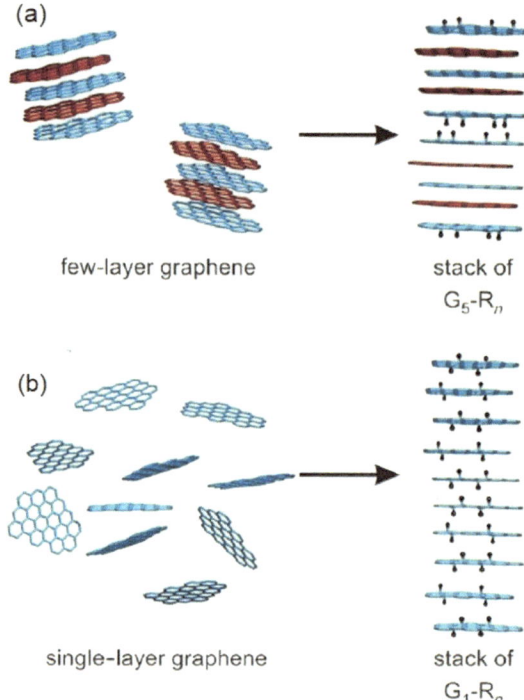

Fig. 2.18 Schematic picture of **a** five-layer graphene G_5 functionalized with n R groups (G_5–R_n) and **b** similarly functionalized single-layer graphene G_1 to give stacks of G_1–R_n (reused with permissions from [152] Copyright ©2014 Wiley-VCH Verlag GmbH & Co. KGaA, Weinheim)

Fig. 2.19 Structure of graphane in the chair conformation. The carbon atoms are shown in gray and the hydrogen atoms in white. While maintaining the graphene hexagonal symmetry, graphane has altered electronic properties and local structure since the attachment of hydrogen atoms to sp^2 carbons changes their hybridization state to sp^3. (reused with permissions from [153] Copyright ©2007 American Physical Society)

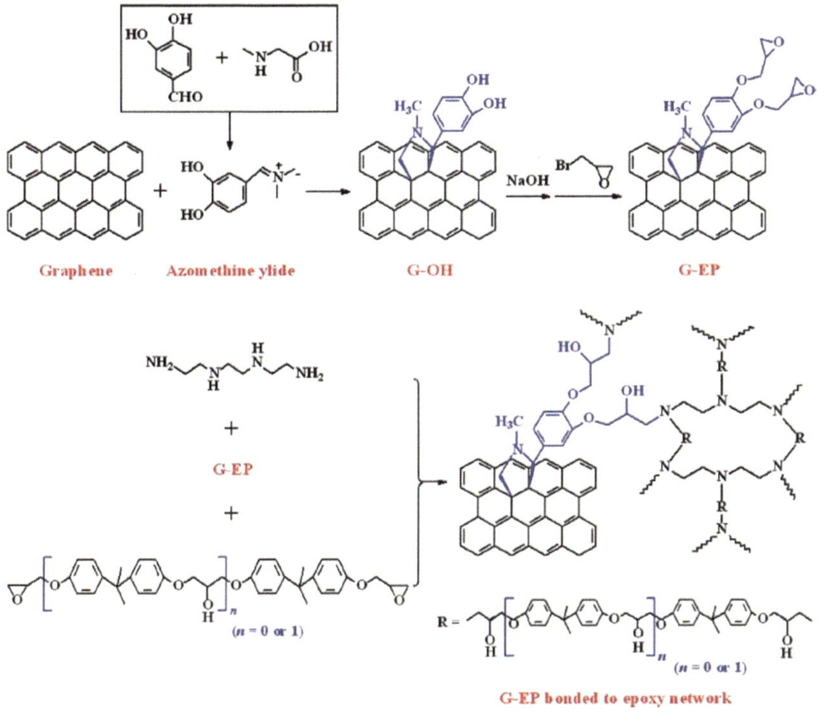

Fig. 2.20 Synthesis of the azomethine ylide-functionalized graphene (G-OH) and epoxide-functionalized graphene (G-EP) (reused with permissions from [156] Copyright ©2016 Elsevier Ltd.)

schematically shown in the upper part of Fig. 2.20, each of its DHPP units includes three categories totaling six reactive sites, hereby making G-OH a promising platform to implement the covalent functionalization of graphene with high efficiency. Meanwhile, owing to the controllable feature of 1,3-dipolar cycloaddition used for the synthesis of G-OH, G-OH and its derivatives can succeed in large part in the electronic structure and properties of pristine graphene. To exemplify that G-OH is a competent platform, four graphene-derived materials, i.e., graphene-FeCl$_3$ complex (G-FeCl$_3$), phenolic resin-functionalized graphene (G-PR), epoxide-functionalized graphene (G-EP), and methacryloyl-functionalized graphene (G-MA), have been prepared from G-OH [157].

Modification of graphene with polymers has been attempted using two main strategies. The first consists of coupling graphene and polymers through a simple chemical reaction, known as the *grafting-to* method. The second approach called the *grafting-from* uses graphene as the initiation site to grow a polymer [98, 158]. These methods use either pure or oxidized graphene when initiator sites are fixed from where the polymerization takes place. These sites are fixed both to the basal planes and the edges of graphene by simple chemical reactions, similar to those employed in acid-defect group chemistry and sidewall modification used in the case of carbon nanotubes (CNTs). However, it should be emphasized that GO is significantly more versatile than both graphene and CNTs because a wide range of chemical reactions can be undertaken on its surface due to the abundance of different oxygenated species that can be present, each with their specific chemical reactivity [84]. As the polymer grows from the graphene, and hence away from the surface, the steric effect is minimized, which is very different from that which occurs when the graphene forms part of the main chain. This constitutes the main advantage of the grafting-from method.

In the grafting-from methods, the specific initiators on the graphitic sheets should be immobilized. This may not be possible in certain cases, where the covalent linkage between the pre-synthesized polymer and graphene emerges as the only alternative, expanding the type of polymers that can be bound to graphene. To achieve this purpose, two possibilities exist that form two grafting-to strategies: either the graphene with the adequate functional groups required to react with specific polymers has to be provided, or the polymer with functionalities to react with graphene or GO has to be used.

In the grafting-to method, the polymer chains are first synthesized and, finally, these pre-synthesized polymers are appended with the functional groups of GO or rGO or with its aromatic surface. The simple techniques of the grafting-to method are the direct covalent linkage of the functional polymers on the GO surface using esterification, amidation, click chemistry, nitrene chemistry, radical addition, etc. Covalent binding of pre-synthesized and end-functionalized polymer chains takes place to the surface of suitably functionalized graphene sheets. The benefit of pre-synthesized polymer is that one can synthesize it with the help of a controlled living polymerization technique outside. However, during attachment with the graphene surface, there is a possibility of low graft density arising mainly from the steric hindrance, and it is very difficult to control the graft density in the grafting-to technique. Further, the attachment of a pre-synthesized polymer chain on a graphene surface usually requires long reaction times because of the low diffusion constant of the polymer [98].

The rate-controlling step of propagation is associated with the diffusion of monomers to the chain ends as the chains grow from the surface of graphene and therefore yield a well-defined brush-like structure with high grafting density. Thus, the grafting-from approach overcomes the low grafting density and slow reactivity problem associated with the grafting-to approach. Another advantage of the grafting-from approach is the possibility of forming thinner graphene sheets as the initiation starts from the initiator attached to the graphene surface by attachment of the monomer followed by propagation. During the chain propagation, the interlayer spacing of graphene may gradually increase with the growing chain's size, resulting in the generation of thin graphene sheets by detachment from the stacked graphene sheet. In the grafting-to approach, this is a quite difficult task. So the grafting-from technique may be a better approach for graphene modification, particularly for the quick formation of high graft density and processable thin graphene sheets.

In the work of Rubio et al. [159], poly(methyl methacrylate) (PMMA) with vary-ing molecular weights was covalently attached to the GNP layers using both methods (Fig. 2.21). The exfoliation of the GNP starting material was performed using a stan-dard methodology developed for grafting short alkyl groups: sodium and naphthalene were used as the reducing agent and transfer reagent, respectively. Tetrahydrofuran (THF) was used as the solvent due to its ability to coordinate sodium ions. PMMA was grafted from the graphene by adding methyl methacrylate (MMA) monomer to the chemically reduced graphene solution. The grafting ratios were higher (44.6% to 126.5%) for the grafting-from approach compared to the grafting-to approach (12.6% to 20.3%).

Among different types of grafting-from polymerization, atom transfer radical polymerization (ATRP) has been most widely used [160–166, 183]. ATRP is advan-tageous for radical polymerization that provides a fast initiation process and the development of a dynamic equilibrium between dormant and growing radicals [168]. Moreover, a wide range of monomers with controlled chain lengths can be polymer-ized by ATRP, and block copolymers can be prepared because of the living radical process. Furthermore, ATRP is of the most practical interest in preparing functional polymers as the terminal alkyl halide can be converted to a wide variety of function-alities by using conventional organic synthetic procedures.

Other approaches include polycondensation [169], ring-opening polymeriza-tion [170], reversible addition-fragmentation chain transfer (RAFT) mediated mini-emulsion polymerization [171], direct electrophilic substitution [172], and Ziegler–Natta polymerization [173]. Several studies of polymer-modified graphene by the grafting method are summarized in Tables 2.2 and 2.3.

In the last decades, the chemical functionalization of other carbon allotropes, such as fullerenes and carbon nanotubes, has attracted considerable attention [152]. Numerous covalent and noncovalent derivatives of fullerenes and nanotubes exhibit outstanding properties. Most generally, it can be thought that graphenes would exhibit chemical properties very much reminiscent of those of fullerenes and carbon nan-otubes, especially for addition reactions to the conjugated π system. However, in

Fig. 2.21 Schematic picture distinguishing between the grafting-from and grafting-to methods used to bind polymers to graphene (reused with permissions from [159] Copyright ©2017 American Chemical Society)

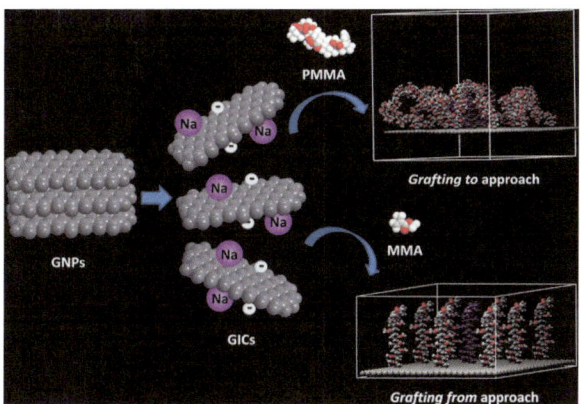

Table 2.2 Methods for grafting polymers from graphene and its derivatives by using the grafting-from technique [158]

Polymer	Form of graphene	Type of polymerization	Notable findings	References
PS, PtBA, PMMA	GO	ATRP	Enhanced solubility, mechanically flexible macroporous carbon films	[160, 161]
PMMA	GO	ATRP	Better solubility, use as filler of pure PMMA, improved mechanical properties	[163]
PDMAEMA	GO	ATRP	Improved solubility	[165]
PS	rGO	ATRP	Enhanced mechanical properties, control of grafting density	[164, 183]
PEMA	rGO	ATRP	Enhanced diffraction efficiency	[162]
PtBA	GO	ATRP	Integration into an electro-active polymer matrix	[166]
PMMA	GO	Radical	Improved thermal and mechanical properties	[174]
PU	rGO	Condensation	Remarkable thermal and mechanical properties	[169]
PCL	GO	Ring-opening	Improved mechanical properties	[170]
PS	GO	RAFT	Improved mechanical and thermal properties	[171]
p-PEK	Graphite	Electrophilic	Substitution	[172]
PP	GO	Ziegler–Natta	Enhanced electrical conductivity	[173]

Table 2.3 Reactions used for grafting polymers to graphene and its derivatives (grafting-to) [158, 175]

Polymer	Form of graphene	Type of reaction	Notable findings	References
PS	fGO	Radical addition	Improvement of PS thermal stability	[176]
PHPMA	fGO	RAFT	Dispersibility in organic solvents and aqueous media	[177]
PS	fGO	Radical addition	Facile and clean method for polymer-functionalization of GO	[178]
PVA	GO	Esterification		[179]
PEI	rGO	Amidation	Formation of hybrid carbon films for supercaps	[180]
PVC	GO, rGO	Esterification	Better mechanical properties	[181]
P3HT	GO	Esterification	Enhanced power conversion efficiency	[182]
Epoxy	GO	Cross-linking by amine-induced ring opening	Improved mechanical properties	[183]
Epoxy	Graphene	Cross-linking by amine-induced ring opening		[184]
PNIPAM	GO	Cross-linking by esterification/nucleophilic substitution	Thermal and pH responsive hydrogels	[185]
PNIPAM	GO	ATNRC	Improved dispersability in water and organic solvents	[186]
PMMA	rGO	Radical grafting	Electrical conductivity	[187]
PS	GO	Click	Solubility in organic media	[188]
PNIPAM	GO	Click	Drugs delivery	[189]
NYLON	GO	Condensation	Better mechanical properties	[190]

(continued)

Table 2.3 (continued)

Polymer	Form of graphene	Type of reaction	Notable findings	References
PNIPAM	Graphene	$\pi - \pi$ interaction RAFT	Application of thermo-responsive composite as a LED light switch to thermo-controlling the light on/off	[191]
PMMA-b-PDMS	GO	$\pi - \pi$ interaction ARGET ATRP	Improvement of mechanical, optical, and thermal properties of the matrix	[192]
PTi-g-PMMA	rGO	$\pi - \pi$ interaction ATRP	Superior mechanical and electronic properties of the composite	[193]

contrast to fullerenes and carbon nanotubes, graphene is a flat and strain-free system whose plane can be functionalized from both sides when dispersed in a solvent.

In natural graphite, the graphene layers stick together through very pronounced $\pi - \pi$ stacking interactions Hence, the wet chemistry of graphene is always concerned with overcoming these noncovalent interlayer binding interactions. For example, an exfoliation of graphite or the stabilization of solvent-dispersed graphene sheets always competes with reaggregation. It is therefore surmised that a solid sample of graphene can be stabilized either on a substrate support or by covering the surface through chemical functionalization.

In general, noncovalent chemistry is attractive because of the preservation of the conjugated π system. The noncovalent functionalization is based on weak interactions between graphene and a binding partner such as a surfactant ligand. Graphene derived from GO can also be combined with surfactants for stabilization [194]. For the covalent functionalization of graphene, a covalent bond must be formed, which is accompanied by the rehybridization of C atoms from sp^2 to sp^3. Although C–O bonds are formed during the synthesis of GO, C–C bonds can be formed, e.g., through the use of diazonium compounds [152]. The drawback of covalent functionalization is that the perfect structure of graphene is destroyed, resulting in significant changes in its physical properties.

Noncovalent interactions primarily involve hydrophobic, Van der Waals, and electrostatic forces and require the physical adsorption of suitable molecules on the graphene surface. Noncovalent functionalization is achieved by polymer wrapping, adsorption of surfactants or small aromatic molecules, and interaction with porphyrins or biomolecules such as deoxyribonucleic acid (DNA) and peptides [195]. H-bonding and $\pi - \pi$ stacking play an important role in noncovalent functionalization enhancing the solubility and assembly without affecting the $\pi - \pi$ conjugation of the skeleton of graphene [151].

Various surface-active molecules, such as sodium cholate, cetyltrimethylammonium bromide, polyvinylpyrrolidone, triphenylene, or pyrene derivatives have been used to produce noncovalently functionalized graphene [152]. An example is shown in Fig. 2.22.

The most common methods of covalent functionalization of GO are the introduction of nucleophilic species, such as substituted amines or hydroxyls, which produce covalently attached functional groups to the carboxylic acid groups located on GO platelets via the formation of amides or esters. The coupling reactions often require activation of the acid group using thionyl chloride (SOCl$_2$), 1-ethyl-3-(3-dimethylaminopropyl)-carbodiimide (EDC), N,N0-dicyclohexylcarbodiimide (DCC), or 2-(7-aza-1H-benzotriazole-1-yl)-1,1,3,3-tetramethyluronium hexafluorophosphate (HATU). Aside from activation and amidation/esterification of the carboxyls on GO, it is also possible to convert them into other reactive groups. As to the epoxy groups, they can be easily modified through ring-opening reactions under various conditions. In addition to small molecules, polymers have also been attached to the surface of GO. These attachments are typically made by either grafting-to or grafting-from approaches. Grafted polymers afford CMGs with increased dispersibility in many solvents, including water, methanol, and polar aprotic organic solvents, depending on the monomer used [84]. Moreover, the decoration of GO sheets with DNA aptamers, peptides, metal nanoparticles, and star polymer micelles as well as polymer grafting has been shown [197]. A schematic representation of the covalent

Fig. 2.22 **a** Schematic illustration of the graphene exfoliation process. Graphite flakes are combined with sodium cholate (SC) in aqueous solution. Sonication exfoliates FLG flakes encapsulated by SC micelles. **b** Graphene dispersion in SC six weeks after it was prepared. **c** Schematic illustration of an ordered SC monolayer on graphene (reused with permissions from [196] Copyright ©2009 American Chemical Society)

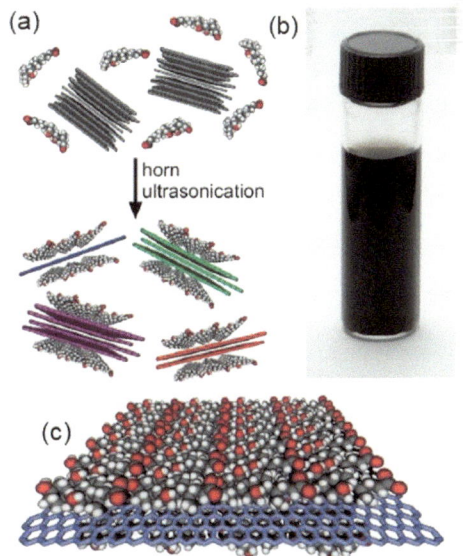

functionalization of GO and subsequent preparation of GO-epoxy nanocomposite is illustrated in Fig. 2.23.

Graphene oxide can also exhibit noncovalent binding via $\pi - \pi$ stacking, cation $-\pi$ or Van der Waals interactions on the sp^2 networks that are not oxidized or engaged in hydrogen bonding. Chemically or thermally reduced GO has been frequently modified by noncovalent physisorption of both polymers and small molecules onto their basal planes via $\pi - \pi$ stacking or Van der Waals interactions [84]. Table 2.4 shows different noncovalent modifications of GO using different modifying agents, their dispersion stability in various solvents, dispersibility, and electrical conductivity.

As a final remark on the above basic consideration, the functionalization of GO creates holes at the carbon basal extension owing to the change in the sp^2 hybridized carbon arrangement of the C layers. In general, nanopores can be introduced into graphene's structure with the unsaturated carbon atoms at the pore edge passivated by chemical functional groups. There is a wide variety of methods for introducing nanopores in graphene with rapid progress in performance. They include electron beam exposure, diblock copolymer templating, helium ion beam drilling, chemical etching, etc. Since 2010, graphene has attracted considerable attention as membranes for producing nanopore devices, owing to their subnanometer thickness that can in theory provide the highest possible spatial resolution of detection [211–213].

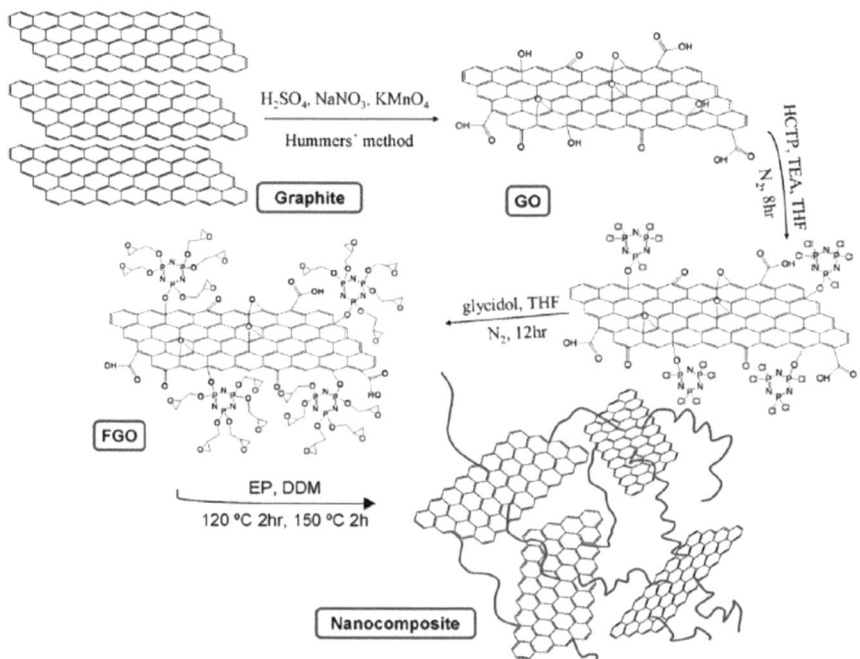

Fig. 2.23 Preparation of functionalized graphene oxide (FGO) and FGO/epoxy nanocomposite (reused with permissions from [198] Copyright ©2011 the Royal Society of Chemistry)

Table 2.4 Noncovalent modification of GO using different modifying agents, their dispersion stability in various solvents, dispersibility, and electrical conductivity [195]

Modifying agent	Dispersing medium	Dispersibility (mg ml^{-1})	Electrical conductivity (S m^{-1})	References
PSS	Water	1		[199]
SPANI	Water	>1	30	[200]
PBA	Water	0.1	200	[201]
Amine terminated polymer	1,3-Dimethyl-2-imidazolidinone, γ-butyrolactone, 1-propanol, ethanol, ethylene glycol, DMF	0.4	1500	[172]
PSSA-g-PPY	Water	3		[202]
Coronene derivative	Water	0.15		[203]
PPESO$_3^-$	Water	0.25	30 kΩ (resistance)	[204]
SDBS	Water	1	80 Ω (resistance)	[205, 206]
MG	Water	0.1		[207]
SLS, SCMC, HPC-Py	Water	0.6–2		[208]
Porphyrin	Water	0.02	370 Ω cm (resistivity)	[209]
PIL	Water	1.5	3600	[210]

Moreover, it can be concluded from the preceding discussion that graphene materials can be electrically conductive, which potentially enables alternative measurement schemes relying on the transverse current across the membrane material itself and thereby extends the technical capability of traditional ionic current-based nanopore devices.

In particular, the arrangement and localization of electrolyte ions during the electrochemical reactions and structuring close to the graphene interface can strongly affect the interface's functional properties and offer a versatile platform for innovative hybrid materials with unique properties [214, 215]. The interfacial behavior of graphene is involved in numerous technological processes and applications, ranging from energy storage to sensing and nanofluidics. The organization of ions and structuring of water molecules close to a graphene interface, which represents an atomically thin surface, substantially affect the interfacial properties in electrolytes as well as the specific capacitance of supercapacitors. Moreover, the adsorption of ions on one side of the ultimately thin material may largely impact the adsorption of additional charge carriers on the opposite side and thus influence the overall supercapacitor performance. However, these phenomena are so far not fully understood.

References

1. Q. Bao, K.P. Loh, Graphene photonics, plasmonics, and broadband optoelectronic devices. ACS Nano **6**, 3677–3694 (2012)
2. J. Li, H. Zeng, Z. Zeng, Y. Zeng, T. Xie, Promising graphene-based nanomaterials and their biomedical applications and potential risks: A comprehensive review. ACS Biomater. Sci. Eng. **7**, 5363–5396 (2021)
3. T. Mahmoudi, Y. Wang, Y.-B. Hahn, Graphene and its derivatives for solar cells application. Nano Energy **47**, 51–65 (2018)
4. C.D. Reddy, S. Rajendran, K.M. Liew, Equilibrium configuration and continuum elastic properties of finite sized graphene. Nanotechnology **17**, 864–870 (2006)
5. P.L. Andres, R. Ramírez, J.A. Vergés, Strong covalent bonding between two graphene layers. Phys. Rev. B **77**, 045403 (2008)
6. A.H. Castro Neto, F. Guinea, N.M.R. Peres, K.S. Novoselov, A.K. Geim, The electronic properties of graphene. Rev. Mod. Phys. **81**, 109–162 (2009)
7. G. Yang, L. Li, W.B. Lee, M.C. Ng, Structure of graphene and its disorders: a review. Sci. Technol. Adv. Mater. **19**, 613–648 (2018)
8. M. Born, K. Huang, *Dynamic Theory of Crystal Lattice* (Clarendon, Oxford, 1954)
9. N.D. Mermin, Crystalline order in two dimensions. Phys. Rev. **176**, 250–254 (1968)
10. N.D. Mermin, H. Wagner, Absence of ferromagnetism or antiferromagnetism in one- or two-dimensional isotropic Heisenberg models. Phys. Rev. Lett. **17**, 1133–1136 (1966)
11. Q. Ai, K. Jarolimek, S. Mazza, J.E. Anthony, C. Risko, Delimited polyacenes: edge topology as a tool to modulate carbon nanoribbon structure, conjugation, and mobility. Chem. Mater. **30**, 947–957 (2018)
12. P. Ruffieux, S. Wang, B. Yang, C. Sánchez-Sánchez, J. Liu, T. Dienel, L. Talirz, P. Shinde, C.A. Pignedoli, D. Passerone, T. Dumslaff, X. Feng, K. Müllen, R. Fasel, On-surface synthesis of graphene nanoribbons with zigzag edge topology. Nature **531**, 489–492 (2016)
13. S. Deng, V. Berry, Wrinkled, rippled and crumpled graphene: an overview of formation mechanism, electronic properties, and applications. Mater. Today. **19**, 197–212 (2016)
14. J.C. Meyer, A.K. Geim, M.I. Katsnelson, K.S. Novoselov, T.J. Boo, S. Roth, The structure of suspended graphene sheets. Nature **446**, 60–63 (2007)
15. W. Bao, F. Miao, Z. Chen, H. Zhang, W. Jang, C. Dames, C.N. Lau, Controlled ripple texturing of suspended graphene and ultrathin graphite membranes. Nat. Nanotechnol. **4**, 562–566 (2009)
16. J. Zang, C. Cao, Y. Feng, J. Liu, X. Zhao, Stretchable and high-performance supercapacitors with crumpled graphene papers. Sci. Rep. **4**, 6492 (2014)
17. A. Fasolino, J. Los, M.I. Katsnelson, Intrinsic ripples in graphene. Nat. Mater. **6**, 858–861 (2007)
18. V.B. Shenoy, C.D. Reddy, A. Ramasubramaniam, Y. Zhang, Edge-stress-induced warping of graphene sheets and nanoribbons. Phys. Rev. Lett. **101**, 245501 (2008)
19. J.M. Carlsson, Buckle or break. Nat. Mater. **6**, 801–802 (2007)
20. E. Cerda, L. Mahadevan, Geometry and physics of wrinkling. Phys. Rev. Lett. **90**, 074302 (2003)
21. K.V. Zakharchenko, M.I. Katsnelson, A. Fasolino, Finite temperature lattice properties of graphene beyond the quasiharmonic approximation. Phys. Rev. Lett. **102**, 046808 (2009)
22. F. Scarpa, S. Adhikari, A.S. Phani, Effective elastic mechanical properties of single layer graphene sheets. Nanotechnology **20**, 065709 (2009)
23. Wong, Y.W., Pellegrino, S.: Wrinkled membranes. Part II: analytical models. J. Mech. Mater. Struct. **1**, 25–29 (2006)
24. K. Xu, P. Cao, J.R. Heath, Scanning tunneling microscopy characterization of the electrical properties of wrinkles in exfoliated graphene monolayers. Nano Lett. **9**, 4446–4451 (2009)
25. W. Kuo, N. Tai, T. Chang, Deformation and fracture in graphene nanosheets. Compos. Part A Appl. Sci. Manuf. **51**, 56–61 (2013)

26. H.C. Schniepp, J.-L. Li, M.J. McAllister, H. Sai, M. Herrera-Alonso, D.H. Adamson, R.K. Prud'homme, R. Car, D.A. Saville, I.A. Aksay, Functionalized single graphene sheets derived from splitting graphite oxide. J. Phys. Chem. B **110**, 8535–8539 (2006)

27. M.J. McAllister, J.-L. Li, D.H. Adamson, H.C. Schniepp, A.A. Abdala, J. Liu, M. Herrera-Alonso, D.L. Milius, R. Car, R.K. Prud'homme, I.A. Aksay, Single sheet functionalized graphene by oxidation and thermal expansion of graphite. Chem. Mater. **19**, 4396–4404 (2007)

28. X. Wang, J. Jin, M. Song, An investigation of the mechanism of graphene toughening epoxy. Carbon **65**, 324–333 (2013)

29. F. Banhart, J. Kotakoski, A.V. Krasheninnikov, Structural defects in graphene. ACS Nano **5**, 26–41 (2011)

30. M.D. Bhatt, H. Kim, G. Kim, Various defects in graphene: a review. RSC Adv. **12**, 21520–21547 (2022)

31. J. Lahiri, Y. Lin, P. Bozkurt, I.I. Oleynik, M. Batzill, An extended defect in graphene as a metallic wire. Nat. Nanotechnol. **5**, 326–329 (2010)

32. D.J. Appelhans, Z. Lin, M.T. Lusk, Two-dimensional carbon semiconductor: Density functional theory calculations. Phys. Rev. B **82**, 073410 (2010)

33. M.H. Gass, U. Bangert, A.L. Bleloch, P. Wang, R.R. Nair, A.K. Geim, Free-standing graphene at atomic resolution. Nat. Nanotechnol. **3**, 676–681 (2008)

34. J.C. Meyer, C. Kisielowski, R. Erni, M.D. Rossell, M.F. Crommie, A. Zettl, Direct imaging of lattice atoms and topological defects in graphene membranes. Nano Lett. **8**, 3582–3586 (2008)

35. M.M. Ugeda, I. Brihuega, F. Guinea, J.M. Gómez-Rodríguez, Missing atom as a source of carbon magnetism. Phys. Rev. Lett. **104**, 096804 (2010)

36. Z. Liang, Z. Xu, T. Yan, F. Ding, Atomistic simulation and the mechanism of graphene amorphization under electron irradiation. Nanoscale **6**, 2082–2086 (2014)

37. A.A. El-Barbary, R.H. Telling, C.P. Ewels, M.I. Heggie, P.R. Briddon, Structure and energetics of the vacancy in graphite. Phys. Rev. B **68**, 144107 (2003)

38. J. Kotakoski, A.V. Krasheninnikov, K. Nordlund, Energetics, structure and long-range interaction of vacancy defects on carbon nanotubes: Atomistic simulations. Phys. Rev. B **74**, 245420 (2006)

39. Y.H. Lee, S.G. Kim, D. Tomanek, Catalytic growth of single-wall carbon nanotubes: An ab initio study. Phys. Rev. Lett. **78**, 2393–2396 (1997)

40. P.O. Lehtinen, A.S. Foster, A. Ayuela, A. Krasheninnikov, K. Nordlund, R.M. Nieminen, Magnetic properties and diffusion of adatoms on a graphene sheet. Phys. Rev. Lett. **91**, 017202 (2003)

41. O.V. Yazyev, S.G. Louie, Topological defects in graphene: Dislocations and grain boundaries. Phys. Rev. B **81**, 195420 (2010)

42. L.L. Bonilla, A. Carpio, Driving dislocations in graphene. Science **337**, 161–162 (2012)

43. J.-H. Wong, B.-R. Wu, M.-F. Lin, Strain effect on the electronic properties of single layer and bilayer graphene. J. Phys. Chem. C **116**, 8271–8277 (2012)

44. B.W. Jeong, J. Ihm, G.-D. Lee, Stability of dislocation defect with two pentagon-heptagon pairs in graphene. Phys. Rev. B **78**, 165403 (2008)

45. A. Hashimoto, K. Suenaga, A. Gloter, K. Urita, S. Iijima, Direct evidence for atomic defects in graphene layers. Nature **430**, 870–873 (2004)

46. A. Zandiatashbar, G.-H. Lee, S.J. An, S. Lee, N. Mathew, M. Terrones, T. Hayashi, C.R. Picu, J. Hone, N. Koratkar, Effect of defects on the intrinsic strength and stiffness of graphene. Nat. Commun. **5**, 3186 (2014)

47. G.-D. Lee, E. Yoon, K. He, A.W. Robertson, J.H. Warner, Detailed formation processes of stable dislocations in graphene. Nanoscale **6**, 14836–14844 (2014)

48. G. López-Polín, C. Gómez-Navarro, V. Parente, F. Guinea, M.I. Katsnelson, F. Perez-Murano, Gómez-Herrero: Increasing the elastic modulus of graphene by controlled defect creation. Nat. Phys. **11**, 26–31 (2015)

49. O. Leenaerts, B. Partoens, F.M. Peeters, Adsorption of H_2O, NH_3, CO, NO_2, and NO on graphene: A first-principles study. Phys. Rev. B **77**, 125416 (2008)

50. X. Lin, J. Ni, C. Fang, Adsorption capacity of H_2O, NH_3, CO, and NO_2 on the pristine graphene. J. Appl. Phys. **113**, 034306 (2013)
51. L. Kong, A. Enders, T.S. Rahman, P.A. Dowben, Molecular adsorption on graphene. J. Phys. Cond. Matt. **26**, 443001 (2014)
52. O. Jakšíc, M. Spasenovíc, Z. Jakšíc, D. Vasiljevíc-Radovíc, Monolayer gas adsorption on graphene-based materials: Surface density of adsorption sites and adsorption capacity. Surf. **3**, 423–432 (2020)
53. A. Kozbial, F. Zhou, Z. Li, H. Liu, L. Li, Are graphitic surfaces hydrophobic? Acc. Chem. Res. **49**, 2765–2773 (2016)
54. V.M. Miriyala, R. Lo, S. Haldar, A. Sarmah, P. Hobza, Structure and properties of double-sandwich complexes at the graphene surface: A theoretical study. J. Phys. Chem. C **123**, 14712–14724 (2016)
55. M. Kayanuma, M. Shoji, K. Furuya, K. Kamiya, Y. Aikawa, M. Umemura, Y. Shigeta, First-principles study of the reaction mechanism of CHO + H on graphene surface. J. Phys. Chem. A **123**, 5633–5639 (2019)
56. H. Zhu, Y. Xu, Y. Yan, J. Xu, C. Yang, Interfacial diffusion of hydrated ion on graphene surface: A molecular simulation study. Langmuir **36**, 13613–13620 (2020)
57. D.J. Joshi, J.R. Koduru, N.I. Malek, C.M. Hussain, S.K. Kailasa, Surface modifications and analytical applications of graphene oxide: A review. Trends Anal. Chem. **144**, 116448 (2021)
58. A. Shrestha, Y. Sumiya, K. Okazawa, T. Uwabe, K. Yoshizawa, Molecular understanding of adhesion of epoxy resin to graphene and graphene oxide surfaces in terms of orbital interactions. Langmuir **39**, 5514–5526 (2023)
59. W.A. Gill, M.R.S.A. Janjua, Exploring the adsorption mechanism of N_2O on graphene: A DFT study on circum-coronene for catalysis, sensing, and energy storage applications. J. Phys. Chem. A **127**, 5591–5601 (2023)
60. C. Wetzl, A. Silvestri, M. Garrido, Dr. H.-L. Hou, A. Criado, M. Prato, The covalent functionalization of surface-supported graphene: An update. Angew. Chem. **62**, e202212857 (2023)
61. Z. Cao, V. Quintano, R. Joshi, Covalent functionalization of graphene oxide. Carbon **2**, 199–205 (2023)
62. J. Kong, L.S. Levitov, D. Halbertal, E. Zeldov, Resonant electron-lattice cooling in graphene. Phys. Rev. B **97**, 245416 (2018)
63. C.H. Wong, Z. Sofer, M. Kubešová, J. Kučera, S. Matějková, M. Pumera, Synthetic routes contaminate graphene materials with a whole spectrum of unanticipated metallic elements. PProc. Natl. Acad. Sci. U.S.A. **111**, 13774–13779 (2014)
64. I.I. Barbolina, C.R. Woods, N. Lozano, K. Kostarelos, K.S. Novoselov, I.S. Roberts, Purity of graphene oxide determines its antibacterial activity. 2D Mater. **3**, 025025 (2016)
65. R. Jalili, D. Esrafilzadeh, S.H. Aboutalebi, Y.M. Sabri, A.E. Kandjani, S.K. Bhargava, E. Della Gaspera, T.R. Gengenbach, A. Walker, Y. Chao, C. Wang, H. Alimadadi, D.R.G. Mitchell, D.L. Officer, D.R. MacFarlane, G.G. Wallace, Silicon as a ubiquitous contaminant in graphene derivatives with significant impact on device performance. Nat. Commun. **9**, 5070 (2018)
66. D. Huertas-Hernando, F. Guinea, A. Brataas, Spin-orbit coupling in curved graphene, fullerenes, nanotubes, and nanotube caps. Phys. Rev. B **74**, 155426 (2006)
67. T.O. Wehling, M.I. Katsnelson, A.I. Lichtenstein, Impurities on graphene: Midgap states and migration barriers. Phys. Rev. B **80**, 085428 (2009)
68. A. Bianco, H.-M. Cheng, T. Enoki, Y. Gogotsi, R.H. Hurt, N. Koratkar, T. Kyotani, M. Monthioux, C.R. Park, J.M.D. Tascon, J. Zhang, All in the graphene family–A recommended nomenclature for two-dimensional carbon materials. Carbon **65**, 1–6 (2013)
69. A.K. Geim, K.S. Novoselov, The rise of graphene. Nat. Mater. **6**, 183–191 (2007)
70. A.A. Balandin, S. Ghosh, W. Bao, I. Calizo, D. Teweldebrhan, F. Miao, C.N. Lau, Superior thermal conductivity of single-layer graphene. Langmuir **8**, 902–907 (2008)
71. N. Parvin, V. Kumar, S.W. Joo, S.-S. Park, T.K. Mandal, Recent advances in the characterized identification of mono-to-multi-layer graphene and its biomedical applications: A review. Electronics **11**, 3345 (2022)

72. S. Sharif, K.S. Ahmad, F. Rehman, Z. Bhatti, K.H. Thebo, Two-dimensional graphene oxide based membranes for ionic and molecular separation: Current status and challenges. J. Environ. Chem. Eng. **9**, 105605 (2021)

73. R. Ikram, B.M. Jan, W. Ahmad, Advances in synthesis of graphene derivatives using industrial wastes precursors; prospects and challenges. J. Mater. Res. Technol. **9**, 15924–15951 (2020)

74. N.K. Rotte, V. Naresh, S. Muduli, V. Reddy, V.V.S. Srikanth, S.K. Martha, Microwave aided scalable synthesis of sulfur, nitrogen co-doped few-layered graphene material for high-performance supercapacitors. Electrochim. Acta **363**, 137209 (2020)

75. R. Sekiya, T. Haino, Edge-functionalized nanographenes. Chem. **27**, 187–199 (2021)

76. A.T. Dideikin, A.Y. Vul', Graphene oxide and derivatives: The place in graphene family. Front. Phys. **6**, 149 (2019)

77. W. Gao, L.B. Alemany, L. Ci, P.M. Ajayan, New insights into the structure and reduction of graphite oxide. Nat. Chem. **1**, 403–408 (2009)

78. K.A. Mkhoyan, A.W. Contryman, J. Silcox, D.A. Stewart, G. Eda, C. Mattevi, S. Miller, M. Chhowalla, Atomic and electronic structure of graphene-oxide. Nano Lett. **9**, 1058–1063 (2009)

79. L. Sun, Structure and synthesis of graphene oxide. Chin. J. Chem. Eng. **27**, 2251–2260 (2019)

80. W. Yu, L. Sisi, Y. Haiyana, L. Jie, Progress in the functional modification of graphene/graphene oxide: a review. RSC Adv. **10**, 15328–15345 (2020)

81. W. Tao, Y. Lan, J. Zhang, L. Zhu, Q. Liu, Y. Yang, S. Yang, G. Tian, S. Zhang, Revealing the chemical nature of functional groups on graphene oxide by integrating potentiometric titration and ab initio calculations. ACS Omega **8**, 24332–24340 (2023)

82. A. Lerf, H. He, M. Forster, J. Klinowski, Structure of graphite oxide revisited. J. Phys. Chem. B **102**, 4477–4482 (1998)

83. T. Szabó, O. Berkesi, P. Forgó, K. Josepovits, Y. Sanakis, D. Petridis, I. Dékány, Evolution of surface functional groups in a series of progressively oxidized graphite oxides. Chem. Mater. **18**, 2740–2749 (2006)

84. D.R. Dreyer, S. Park, C.W. Bielawski, R.S. Ruoff, The chemistry of graphene oxide. Chem. Soc. Rev. **39**, 228–240 (2010)

85. S.H. Aboutalebi, M.M. Gudarzi, Q. Zheng, J.K. Kim, Spontaneous formation of liquid crystals in ultralarge graphene oxide dispersions. Adv. Funct. Mater. **21**, 2978–2988 (2011)

86. S. Sadeghi, A.Z. Yazdi, U. Sundararaj, Controlling short-range interactions by tuning surface chemistry in HDPE/graphene nanoribbon nanocomposites. J. Phys. Chem. B **119**, 11867–11878 (2015)

87. U. Hofmann, R. Holst, Über die säurenatur und die methylierung von graphitoxyd. Ber. Dtsch. Chem. Ges. **72**, 754–771 (1939)

88. G. Ruess, Über das graphitoxyhydroxyd (graphitoxyd). Monatsh. Chem. Verw. Teile Anderer Wiss. **76**, 381–417 (1947)

89. W. Scholz, H.-P.Z. Boehm, Formation process and structure of graphite oxide. Anorg. Allg. Chem. **369**, 327–340 (1969)

90. T. Nakajima, Y. Matsuo, Formation process and structure of graphite oxide. Carbon **32**, 469–475 (1994)

91. S. Eigler, A. Hirsch, Chemistry with graphene and graphene oxide-challenges for synthetic chemists. Langmuir **49**, 2765–2773 (2023)

92. J.P. Rourke, P.A. Pandey, J.J. Moore, M. Bates, I.A. Kinloch, R.J. Young, N.R. Wilson, The real graphene oxide revealed: stripping the oxidative debris from the graphene-like sheets. Angew. Chem. **123**, 3231–3235 (2011)

93. C.H. Lui, L. Liu, K.F. Mak, G.W. Flynn, T.F. Heinz, Ultraflat graphene. Nature **462**, 339–341 (2009)

94. J.I. Paredes, S. Villar-Rodil, P. Solís-Fernández, A. Martínez-Alonso, J.M.D. Tascón, Atomic force and scanning tunneling microscopy imaging of graphene nanosheets derived from graphite oxide. Langmuir **25**, 5957–5968 (2009)

95. C. Gómez-Navarro, J.C. Meyer, R.S. Sundaram, A. Chuvilin, S. Kurasch, M. Burghard, K. Kern, U. Kaiser, Atomic structure of reduced graphene oxide. Nano Lett. **10**, 1144–1148 (2010)

96. D. Pandey, R. Reifenberger, R. Piner, Scanning probe microscopy study of exfoliated oxidized graphene sheets. Surf. Sci. **602**, 1607–1613 (2008)
97. V. Georgakilas, J.N. Tiwari, K.C. Kemp, J.A. Perman, A.B. Bourlinos, K.S. Kim, R. Zboril, Noncovalent functionalization of graphene and graphene oxide for energy materials, biosensing, catalytic, and biomedical applications. Chem. Rev. **116**, 5464–5519 (2016)
98. R.K. Layek, A.K. Nandi, A review on synthesis and properties of polymer functionalized graphene. Polymer **54**, 5087–5103 (2013)
99. P. Katti, K.V. Kundan, S. Kumar, S. Bose, Improved mechanical properties through engineering the interface by poly (ether ether ketone) grafted graphene oxide in epoxy based nanocomposites. Polymer **122**, 184–193 (2017)
100. A.B. Kaiser, Electronic transport properties of conducting polymers and carbon nanotubes. Rep. Prog. Phys. **64**, 1–49 (2001)
101. Y. Kopelevich, P. Esquinazi, Graphene physics in graphite. Adv. Mater. **19**, 4559–4563 (2007)
102. H.A. Becerril, J. Mao, Z. Liu, R.M. Stoltenberg, Z. Bao, Y. Chen, Evaluation of solution-processed reduced graphene oxide films as transparent conductors. ACS Nano **2**, 463–470 (2008)
103. J. Zhao, S. Pei, W. Ren, L. Gao, H.-M. Cheng, Efficient preparation of large-area graphene oxide sheets for transparent conductive films. ACS Nano **4**, 5245–5252 (2010)
104. S. Stankovich, D.A. Dikin, R.D. Piner, K.A. Kohlhaas, A. Kleinhammes, Y. Jia, Y. Wu, S.T. Nguyen, R.S. Ruoff, Synthesis of graphene-based nanosheets via chemical reduction of exfoliated graphite oxide. Carbon **45**, 1558–1565 (2007)
105. S. Pei, H. Cheng, The reduction of graphene oxide. Carbon **50**, 3210–3228 (2012)
106. S. Gilje, S. Han, M. Wang, K.L. Wang, R.B. Kaner, A chemical route to graphene for device applications. Nano Lett. **7**, 3394–3398 (2007)
107. S. Watcharotone, D.A. Dikin, S. Stankovich, R. Piner, I. Jung, G.H.B. Dommett, G. Evmenenko, S. Wu, S. Chen, C. Liu, S.T. Nguyen, R.S. Ruoff, Graphene-silica composite thin films as transparent conductors. Nano Lett. **7**, 1888–1892 (2007)
108. K.S. Novoselov, A.K. Geim, S.V. Morozov, D. Jiang, Y. Zhang, S.V. Dubonos, I.V. Grigorieva, A.A. Firsov, Electric field effect in atomically thin carbon films. Science **306**, 666–669 (2004)
109. S. Bae, H. Kim, Y. Lee, X. Xu, J.-S. Park, Y. Zheng, J. Balakrishnan, T. Lei, H.R. Kim, Y.I. Song, Y.-J. Kim, K.S. Kim, B. Özyilmaz, J.-H. Ahn, B.H. Hong, S. Iijima, Roll-to-roll production of 30-inch graphene films for transparent electrodes. Nat. Nanotechnol. **5**, 574–578 (2010)
110. G. Eda, M. Chhowalla, Chemically derived graphene oxide: towards large-area thin-film electronics and optoelectronics. Adv. Mater. **22**, 2392–2415 (2010)
111. G.B. Olowojoba, S. Kopsidas, S. Eslava, E.S. Gutierrez, A.J. Kinloch, C. Mattevi, V.G. Rocha, A.C. Taylor, A facile way to produce epoxy nanocomposites having excellent thermal conductivity with low contents of reduced graphene oxide. J. Mater. Sci. **52**, 7323–7344 (2017)
112. K. Ai, Y. Liu, L. Lu, X. Cheng, L. Huo, A novel strategy for making soluble reduced graphene oxide sheets cheaply by adopting an endogenous reducing agent. J. Mater. Chem. **21**, 3365–3370 (2011)
113. X. Cao, D. Qi, S. Yin, J. Bu, F. Li, C.F. Goh, S. Zhang, X. Chen, Ambient fabrication of large-area graphene films via a synchronous reduction and assembly strategy. Adv. Mater. **25**, 2957–2962 (2013)
114. C. Zhang, W. Lv, W. Zhang, X. Zheng, M. Wu, W. Wei, Y. Tao, Z. Li, Q. Yang, Reduction of graphene oxide by hydrogen sulfide: a promising strategy for pollutant control and as an electrode for Li-S batteries. Adv. Energy Mater. **4**, 1301565 (2014)
115. S.K. Tiwari, V. Kumar, A. Huczko, R. Oraon, A.D. Adhikari, G. Nayak, Magical allotropes of carbon: prospects and applications. Crit. Rev. Solid State Mater. Sci. **41**, 257–317 (2016)
116. D. Hou, Q. Liu, X. Wang, Y. Quan, Z. Qiao, L. Yu, S. Ding, Facile synthesis of graphene via reduction of graphene oxide by artemisinin in ethanol. J. Materiomics **4**, 256–265 (2018)
117. S. Baradaran, E. Moghaddam, W.J. Basirun, M. Mehrali, M. Sookhakian, M. Hamdi, M.N. Moghaddam, Y. Alias, Mechanical properties and biomedical applications of a nanotube hydroxyapatite-reduced graphene oxide composite. Carbon **69**, 32–45 (2014)

118. X. Duan, H. Sun, Z. Ao, L. Zhou, G. Wang, S. Wang, Unveiling the active sites of graphene-catalyzed peroxymonosulfate activation. Carbon **107**, 371–378 (2016)
119. L. Yang, D. Liu, J. Huang, T. You, Simultaneous determination of dopamine, ascorbic acid and uric acid at electrochemically reduced graphene oxide modified electrode. Sens. Actuators B Chem. **193**, 166–172 (2014)
120. S.Y. Toh, K.S. Loh, S.K. Kamarudin, W.R.W. Daud, Graphene production via electrochemical reduction of graphene oxide: synthesis and characterisation. Chem. Eng. J. **251**, 422–434 (2014)
121. H.Q. Sun, S.Z. Liu, G.L. Zhou, H.M. Ang, M.O. Tade, S.B. Wang, Reduced graphene oxide for catalytic oxidation of aqueous organic pollutants. ACS Appl. Mater. Interfaces **4**, 5466–5471 (2012)
122. Y. Su, V. Kravets, S. Wong, J. Waters, A. Geim, R. Nair, Impermeable barrier films and protective coatings based on reduced graphene oxide. Nat. Commun. **5**, 4843–2773 (2014)
123. S.K. Tiwari, R.K. Mishra, S.K. Ha, A. Huczko, Evolution of graphene oxide and graphene: From imagination to industrialization. ChemNanoMat **4**, 598–620 (2018)
124. P. Cataldi, A. Athanassiou, I.S. Bayer, Graphene nanoplatelets-based advanced materials and recent progress in sustainable applications. Appl. Sci. **8**, 1438 (2018)
125. P. Wick, A.E. Louw-Gaume, M. Kucki, H.F. Krug, K. Kostarelos, B. Fadeel, K.A. Dawson, A. Salvati, E. Vázquez, L. Ballerini, M. Tretiach, F. Benfenati, E. Flahaut, L. Gauthier, M. Prato, A. Bianco, Classification framework for graphene-based materials. Angew. Chem. Int. Ed. **53**, 7714–7718 (2014)
126. B.Z. Jang, A. Zhamu, Processing of nanographene platelets (NGPs) and NGP nanocomposites: a review. J. Mater. Sci. **43**, 5092–5101 (2008)
127. R. Sengupta, M. Bhattacharya, S. Bandyopadhyay, A.K. Bhowmick, A review on the mechanical and electrical properties of graphite and modified graphite reinforced polymer composites. Prog. Polym. Sci. **36**, 638–670 (2011)
128. R.J. Young, I.A. Kinloch, L. Gong, K.S. Novoselov, The mechanics of graphene nanocomposites: A review. Compos. Sci. Technol. **72**, 1459–1476 (2012)
129. A.E.D.R. Castillo, V. Pellegrini, A. Ansaldo, F. Ricciardella, H. Sun, L. Marasco, J. Buha, Z. Dang, L. Gagliani, E. Lago, N. Curreli, S. Gentiluomo, F. Palazon, M. Prato, R. Oropesa-Nuñez, P.S. Toth, E. Mantero, M. Crugliano, A. Gamucci, A. Tomadin, M. Polini, F. Bonaccorso, High-yield production of 2D crystals by wet-jet milling. Mater. Horiz. **5**, 890–904 (2018)
130. S.K. Tiwari, S. Sahoo, N. Wang, A. Huczko, Graphene research and their outputs: Status and prospect. J. Sci. Adv. Mater. Devices **5**, 10–29 (2020)
131. Y. Dong, J. Shao, C. Chen, H. Li, R. Wang, Y. Chi, X. Lin, G. Chen, Blue luminescent graphene quantum dots and graphene oxide prepared by tuning the carbonization degree of citric acid. Carbon **50**, 4738–4743 (2012)
132. H. Sun, L. Wu, W. Wei, X. Qu, Recent advances in graphene quantum dots for sensing. Mater. Today **16**, 433–442 (2013)
133. M.H.M. Facure, R. Schneider, L.A. Mercante, D.S. Correa, A review on graphene quantum dots and their nanocomposites: from laboratory synthesis towards agricultural and environmental applications. Environ. Sci. Nano **7**, 3710–3734 (2020)
134. M. Bacon, S.J. Bradley, T. Nann, Graphene quantum dots. Part. Part. Syst. Charact. **31**, 415–428 (2014)
135. P. Tian, L. Tang, K.S. Teng, S.P. Lau, Graphene quantum dots from chemistry to applications. Mater. Today Chem. **10**, 221–258 (2018)
136. E. Haque, J. Kim, V. Malgras, K.R. Reddy, A.C. Ward, J. You, Y. Bando, M.S.A. Hossain, Y. Yamauchi, Recent advances in graphene quantum dots: Synthesis, properties, and applications. Small Methods **2**, 1800050 (2018)
137. M. Li, T. Chen, J.J. Gooding, J. Liu, Review of carbon and graphene quantum dots for sensing. ACS Sens. **4**, 1732–1748 (2019)
138. S. Zhu, Y. Song, X. Zhao, J. Shao, J. Zhang, B. Yang, The photoluminescence mechanism in carbon dots (graphene quantum dots, carbon nanodots, and polymer dots): current state and future perspective. Nano Res. **8**, 355–381 (2015)

139. S. Kapoor, A. Jha, H. Ahmad, S.S. Islam, Avenue to large-scale production of graphene quantum dots from high-purity graphene sheets using laboratory-grade graphite electrodes. ACS Omega **5**, 18831–18841 (2020)
140. M.W. Lee, J. Kim, J.S. Suh, Characteristics of graphene quantum dots determined by edge structures: three kinds of dots fabricated using thermal plasma jet. RSC Adv. **5**, 67669–67675 (2015)
141. V. Georgakilas, M. Otyepka, A.B. Bourlinos, V. Chandra, N. Kim, K.C. Kemp, P. Hobza, R. Zboril, K.S. Kim, Functionalization of graphene: Covalent and non-covalent approaches, derivatives and applications. Chem. Rev. **112**, 6156–6214 (2012)
142. Q. Tang, Z. Zhou, Z. Chen, Graphene-related nanomaterials: tuning properties by functionalization. Nanoscale **5**, 4541–4682 (2013)
143. M. Yan, Pristine graphene: functionalization, fabrication, and nanocomposite materials. J. Phys. Conf. Ser. **1143**, 012012 (2018)
144. J. Park, M. Yan, Covalent functionalization of graphene with reactive intermediates. Acc. Chem. Res. **46**, 181–189 (2013)
145. S.P. Economopoulos, N. Tagmatarchis, Chemical functionalization of exfoliated graphene. Chem. Eur. J. **19**, 12930–12936 (2013)
146. Z.Y. Liu, D. He, Y.S. Wang, H.P. Wu, J.A. Wang, Solution-processable functionalized graphene in donor/acceptor-type organic photovoltaic cells. Sol. Energy Mater. Sol. Cells **94**, 1196–1200 (2010)
147. T.T. Baby, S.S.J. Aravind, T. Arockiadoss, R.B. Rakhi, S. Ramaprabhu, Metal decorated graphene nanosheets as immobilization matrix for amperometric glucose biosensor. Sens. Actuators B **145**, 71–77 (2010)
148. A. Ito, H. Nakamura, Hydrogen isotope sputtering of graphite by molecular dynamics simulation. Thin Solid Films **516**, 6553–6559 (2008)
149. Z. Cao, X. Wen, V. Quintano, R. Joshi, On the role of functionalization in graphene-moisture interaction. Curr. Opin. Solid State Mater. Sci. **27**, 101122 (2023)
150. K. Kwon, J. Kim, K. Roh, P.J. Kim, J. Choi, Towards high performance Li metal batteries: Surface functionalized graphene separator with improved electrochemical kinetics and stability. Electrochem. commun. **157**, 107598 (2023)
151. V.D. Punetha, S. Rana, H.J. Yoo, A. Chaurasia, J.T. McLeskey, M.S. Ramasamy, N.G. Sahoo, J.W. Cho, Functionalization of carbon nanomaterials for advanced polymer nanocomposites: A comparison study between CNT and graph53ene. Prog. Polym. Sci. **67**, 1–47 (2017)
152. S. Eigler, A. Hirsch, Chemistry with graphene and graphene oxide–challenges for synthetic chemists. Angew. Chem. Int. Ed. **53**, 7720–7738 (2014)
153. J.O. Sofo, A.S. Chaudhari, G.D. Barber, Graphane: A two-dimensional hydrocarbon. Phys. Rev. B **75**, 153401 (2007)
154. D.C. Elias, R.R. Nair, T.M.G. Mohiuddin, S.V. Morozov, P. Blake, M.P. Halsall, A.C. Ferrari, D.W. Boukhvalov, M.I. Katsnelson, A.K. Geim, K.S. Novoselov, Control of graphene's properties by reversible hydrogenation: Evidence for graphane. Science **323**, 610–613 (2009)
155. V. Georgakilas, A.B. Bourlinos, R. Zboril, T.A. Steriotis, P. Dallas, A.K. Stubos, C. Trapalis, Organic functionalisation of graphenes. Chem. Commun. **46**, 1766–1768 (2010)
156. S. Zhao, H. Chang, S. Chen, J. Cui, Y. Yan, High-performance and multifunctional epoxy composites filled with epoxide-functionalized graphene. Eur. Polym. J. **84**, 300–312 (2016)
157. S. Zhao, H. Wang, L. Xin, J. Cui, Y. Yan, A versatile platform of 2-(3,4-dihydroxyphenyl) pyrrolidine grafted graphene for preparation of various graphene-derived materials. Chem. Asian J. **10**, 1177–1183 (2015)
158. H.J. Salavagione, G. Martínez, G. Ellis, Recent advances in the covalent modification of graphene with polymers. Macromol. Rapid Commun. **32**, 1771–1789 (2011)
159. N. Rubio, H. Au, H.S. Leese, S. Hu, A.J. Clancy, M.S.P. Shaffer, Grafting from versus grafting to approaches for the functionalization of graphene nanoplatelets with poly(methyl methacrylate). Macromolecules **50**, 7070–7079 (2017)

160. S.H. Lee, D.R. Dreyer, J. An, A. Velamakanni, R.D. Piner, S. Park, Y. Zhu, S.O. Kim, C.W. Bielawski, R.S. Ruoff, Polymer brushes via controlled, surface-initiated atom transfer radical polymerization (ATRP) from graphene oxide. Macromol. Rapid Commun. **31**, 281–288 (2010)

161. S.H. Lee, H.W. Kim, J.O. Hwang, W.J. Lee, J. Kwon, C.W. Bielawski, R.S. Ruoff, S.O. Kim, Three-dimensional self-assembly of graphene oxide platelets into mechanically flexible macroporous carbon films. Angew. Chem. Int. Ed. **49**, 10084–10088 (2010)

162. D. Wang, G. Ye, X. Wang, X. Wang, Graphene functionalized with azo polymer brushes: Surface-initiated polymerization and photoresponsive properties. Adv. Mater. **23**, 1122–1125 (2011)

163. G. Gonçalves, P.A.A.P. Marques, A. Barros-Timmons, I. Bdkin, M.K. Singh, N. Emami, J. Grácio, Graphene oxide modified with PMMA via ATRP as a reinforcement filler. J. Mater. Chem. **20**, 9927–9934 (2010)

164. M. Fang, K. Wang, H. Lu, Y. Yang, S. Nutt, Covalent polymer functionalization of graphene nanosheets and mechanical properties of composites. J. Mater. Chem. **19**, 7098–7105 (2009)

165. Y. Yang, J. Wang, J. Zhang, J. Liu, X. Yang, H. Zhao, Exfoliated graphite oxide decorated by PDMAEMA chains and polymer particles. Langmuir **25**, 11808–11814 (2009)

166. G.L. Li, G. Liu, M. Li, D. Wan, K.G. Neoh, E.T. Kang, Organo- and water-dispersible graphene oxide-polymer nanosheets for organic electronic memory and gold nanocomposites. J. Phys. Chem. C **114**, 12742–12748 (2010)

167. M. Fang, K. Wang, H. Lu, Y. Yang, S. Nutt, Single-layer graphene nanosheets with controlled grafting of polymer chains. J. Mater. Chem. **20**, 1982–1992 (2010)

168. V. Coessens, T. Pintauer, K. Matyjaszewsky, Functional polymers by atom transfer radical polymerization. Prog. Polym. Sci. **26**, 337–377 (2001)

169. X. Wang, Y. Hu, L. Song, H. Yang, W. Xing, H. Lu, In situ polymerization of graphene nanosheets and polyurethane with enhanced mechanical and thermal properties. J. Mater. Chem. **21**, 4222–4227 (2011)

170. S.M. Kang, S. Park, D. Kim, S.Y. Park, R.S. Ruoff, H. Lee, Simultaneous reduction and surface functionalization of graphene oxide by mussel-inspired chemistry. Adv. Funct. Mater. **21**, 108–112 (2011)

171. H.M. Etmimi, M.P. Tonge, R.D. Sanderson, Synthesis and characterization of polystyrene-graphite nanocomposites via surface RAFT-mediated miniemulsion polymerization. J. Polym. Sci. A Polym. Chem. **49**, 1621–1632 (2011)

172. E.-K. Choi, I.-Y. Jeon, S.J. Oh, J.B. Baek, Direct grafting of linear macromolecular wedges to the edge of pristine graphite to prepare edge-functionalized graphene-based polymer composites. J. Mater. Chem. **20**, 10936–10942 (2010)

173. Y. Huang, Y. Qin, Y. Zhou, H. Niu, Z.-Z. Yu, J.-Y. Dong, Polypropylene/graphene oxide nanocomposites prepared by in situ Ziegler-Natta polymerization. Chem. Mater. **22**, 4096–4102 (2010)

174. K.P. Pramoda, H. Hussain, H.M. Koh, H.R. Tan, C.B. He, Covalent bonded polymer-graphene nanocomposites. J. Polym. Sci. A Polym. Chem. **48**, 4262–4267 (2010)

175. P. Eskandari, Z. Abousalman-Rezvani, H. Roghani-Mamaqani, M. Salami-Kalajahi, H. Mardani, Polymer grafting on graphene layers by controlled radical polymerization. Adv. Colloid Interface Sci. **273**, 102021 (2019)

176. F. Beckert, A.M. Rostas, R. Thomann, S. Weber, E. Schleicher, C. Friedrich, R. Mülhaupt, Self-initiated free radical grafting of styrene homo- and copolymers onto functionalized graphene. Macromolecules **46**, 5488–5496 (2013)

177. Z. Liu, G. Lu, Y. Li, Y. Li, X. Huang, Click synthesis of graphene/poly(N-(2-hydroxypropyl) methacrylamide) nanocomposite via grafting-onto strategy at ambient temperature. RSC Adv. **4**, 60920–60928 (2014)

178. O. García-Valdez, R. Ledezma-Rodríguez, R. Torres-Lubian, L. Yate, E. Saldívar-Guerra, R.F. Ziolo, The "grafting-to" of well-defined polystyrene on graphene oxide via Nitroxide-mediated polymerization. Macromol. Chem. Phys. **217**, 2099–2106 (2016)

179. L.M. Veca, F. Lu, M.J. Meziani, L. Cao, P. Zhang, G. Qi, L. Qu, M. Shrestha, Y.-P. Sun, Polymer functionalization and solubilization of carbon nanosheets. Chem. Commun. **45**, 2565–2567 (2009)

180. D. Yu, L. Dai, Self-assembled graphene/carbon nanotube hybrid films for supercapacitors. J. Phys. Chem. Lett. **1**, 467–470 (2010)

181. H.J. Salavagione, G. Martínez, Importance of covalent linkages in the preparation of effective reduced graphene oxide-poly(vinyl chloride) nanocomposites. Macromolecules **44**, 2685–2692 (2011)

182. D. Yu, Y. Yang, M. Durstock, J.-B. Baek, L. Dai, Soluble P3HT-grafted graphene for efficient bilayer-heterojunction photovoltaic devices. ACS Nano **4**, 5633–5640 (2010)

183. M. Fang, Z. Zhang, J. Li, H. Zhang, H. Lu, Y. Yang, Constructing hierarchically structured interphases for strong and tough epoxy nanocomposites by amine-rich graphene surfaces. J. Mater. Chem. **20**, 9635–9643 (2010)

184. Y. Fu, W.-H. Zhong, Cure kinetics behavior of a functionalized graphitic nanofiber modified epoxy resin. Thermochim. Acta **516**, 58–63 (2011)

185. S. Sun, P. Wu, A one-step strategy for thermal- and pH-responsive graphene oxide interpenetrating polymer hydrogel networks. J. Mater. Chem. **21**, 4095–4097 (2011)

186. Y. Deng, Y. Li, J. Dai, M. Lang, X. Huang, An efficient way to functionalize graphene sheets with presynthesized polymer via ATNRC chemistry. J. Polym. Sci. A Polym. Chem. **49**, 1582–1590 (2011)

187. D. Vuluga, J.M. Thomassin, I. Molenberg, I. Huynen, B. Gilbert, C. Jerome, M. Alexandre, C. Detrembleur, Straightforward synthesis of conductive graphene/polymer nanocomposites from graphite oxide. Chem. Commun. **47**, 2544–2546 (2011)

188. S. Sun, Y. Cao, J. Feng, P. Wu, Click chemistry as a route for the immobilization of well-defined polystyrene onto graphene sheets. J. Mater. Chem. **20**, 5605–5607 (2010)

189. Y. Pan, H. Bao, N.G. Sahoo, T. Wu, L. Li, Water-soluble poly(N-isopropylacrylamide)-graphene sheets synthesized via click chemistry for drug delivery. Adv. Funct. Mater. **21**, 2754–2763 (2011)

190. Z. Xu, C. Gao, In situ polymerization approach to graphene-reinforced nylon-6 composites. Macromolecules **43**, 6716–6723 (2010)

191. D. Jiang, H. Zhu, W. Yang, L. Cui, J. Liu, One-side non-covalent modification of CVD graphene sheet using pyrene-terminated PNIPAAm generated via RAFT polymerization for the fabrication of thermo-responsive actuators. Sens. Actuators B **239**, 193–202 (2017)

192. S. Song, C. Wan, Y. Zhang, Non-covalent functionalization of graphene oxide by pyrene-block copolymers for enhancing physical properties of poly(methyl methacrylate). RSC Adv. **5**, 79947–79955 (2015)

193. N. Maity, A. Mandal, A.K. Nandi, Synergistic interfacial effect of polymer stabilized graphene via non-covalent functionalization in poly(vinylidene fluoride) matrix yielding superior mechanical and electronic properties. Polymer **88**, 79–93 (2016)

194. Q. Su, S. Pang, V. Alijani, C. Li, X. Feng, K. Müllen, Composites of graphene with large aromatic molecules. Adv. Mater. **21**, 3191–3195 (2009)

195. T. Kuila, S. Bose, A.K. Mishra, P. Khanra, N.H. Kim, J.H. Lee, Chemical functionalization of graphene and its applications. Prog. Mater. Sci. **57**, 1061–1105 (2012)

196. A.A. Green, M.C. Hersam, Solution phase production of graphene with controlled thickness via density differentiation. Nano Lett. **9**, 4031–4036 (2009)

197. M. Dehnert, E.-C. Spitzner, F. Beckert, C. Friedrich, R. Magerle, Subsurface imaging of functionalized and polymer-grafted graphene oxide. Macromolecules **49**, 7415–7425 (2016)

198. C. Bao, Y. Guo, L. Song, Y. Kan, X. Qian, Y. Hu, In situ preparation of functionalized graphene oxide/epoxy nanocomposites with effective reinforcements. J. Mater. Chem. **21**, 13290–13298 (2011)

199. S. Stankovich, R.D. Piner, X. Chen, N. Wu, S.T. Nguyen, R.S. Ruoff, Stable aqueous dispersions of graphitic nanoplatelets via the reduction of exfoliated graphite oxide in the presence of poly(sodium 4-styrenesulfonate). J. Mater. Chem. **16**, 155–158 (2006)

200. H. Bai, Y. Xu, L. Zhao, C. Li, G. Shi, Non-covalent functionalization of graphene sheets by sulfonated polyaniline. Chem. Commun. Issue **13**, 1667–1669 (2009)

201. Y. Xu, H. Bai, G. Lu, C. Li, G. Shi, Flexible graphene films via the filtration of water-soluble noncovalent functionalized graphene sheets. J. Am. Chem. Soc. **130**, 5856–5857 (2008)

202. J. Zhang, J. Lei, R. Pan, Y. Xue, H. Ju, Highly sensitive electrocatalytic biosensing of hypoxanthine based on functionalization of graphene sheets with water-soluble conducting graft copolymer. Biosens. Bioelectron. **26**, 371–376 (2010)

203. A. Ghosh, K.V. Rao, S.J. George, C.N.R. Rao, Noncovalent functionalization, exfoliation, and solubilization of graphene in water by employing a fluorescent coronene carboxylate. Chem. Eur. J. **16**, 2700–2704 (2010)

204. H. Yang, Q. Zhang, C. Shan, F. Li, D. Han, L. Niu, Stable, conductive supramolecular composite of graphene sheets with conjugated polyelectrolyte. Langmuir **26**, 6708–6712 (2010)

205. H. Chang, G. Wang, A. Yang, X. Tao, X. Liu, Y. Shen, Z. Zheng, A transparent, flexible, low-temperature, and solution-processible graphene composite electrode. Adv. Funct. Mater. **20**, 2893–2902 (2010)

206. Q. Zeng, J. Cheng, L. Tang, X. Liu, Y. Liu, J. Li, J. Jiang, Self-assembled graphene-enzyme hierarchical nanostructures for electrochemical biosensing. Adv. Funct. Mater. **20**, 3366–3372 (2010)

207. H. Liu, J. Gao, M. Xue, N. Zhu, M. Zhang, T. Cao, Processing of graphene for electrochemical application: Noncovalently functionalize graphene sheets with water-soluble electroactive methylene green. Langmuir **25**, 12006–12010 (2009)

208. Q. Yang, X. Pan, F. Huang, K. Li, Fabrication of high-concentration and stable aqueous suspensions of graphene nanosheets by noncovalent functionalization with lignin and cellulose derivatives. J. Phys. Chem. C **114**, 3811–3816 (2010)

209. J. Geng, H.T. Jung, Porphyrin functionalized graphene sheets in aqueous suspensions: From the preparation of graphene sheets to highly conductive graphene films. J. Phys. Chem. C **114**, 8227–8234 (2010)

210. T.Y. Kim, H. Lee, J.E. Kim, K.S. Suh, Synthesis of phase transferable graphene sheets using ionic liquid polymers. ACS Nano **4**, 1612–1618 (2010)

211. S. Garaj, W. Hubbard, A. Reina, J. Kong, D. Branton, J.A. Golovchenko, Graphene as a subnanometre trans-electrode membrane. Nature **467**, 190–193 (2010)

212. M. Kim, N.S. Safron, E. Han, M.S. Arnold, P. Gopalan, Fabrication and characterization of large-area, semiconducting nanoperforated graphene materials. Nano Lett. **10**, 1125–1131 (2010)

213. H. Qiu, W. Zhou, W. Guo, Nanopores in graphene and other 2D materials: A decade's journey toward sequencing. ACS Nano **15**, 18848–18864 (2021)

214. M. Pykal, M. Langer, B.B. Prudilova, P. Banas, M. Otyepka, Ion interactions across graphene in electrolyte aqueous solutions. J. Phys. Chem. C **3**, 9799–9806 (2019)

215. S.S. Siwal, Q. Zhang, N. Devi, V.K. Thakur, Carbon-based polymer nanocomposite for high-performance energy storage applications. Polymers **12**, 505 (2020)

Chapter 3
Polymer Nanocomposite as a Highly Inhomogeneous and Disordered Medium

Polymer-based composites were proposed as a new paradigm for materials in the 1960s. By dispersing stiff fibers in a polymer matrix, high-performance lightweight composites were developed [1]. Utilization of epoxy composites in the engineering field has tremendously increased in the last decade due to their unique properties such as high strength-to-weight ratio and design flexibility [2–5]. However, some obstacles hinder the application of these materials such as low impact toughness and poor adhesion of hydrophilic reactive groups with polymer matrix. Currently, one of the most challenging research areas of nanotechnology focuses on the inclusion of nanoparticle fillers into polymers to enhance the functional and physical properties of the polymers. Nanofillers can be used to improve the adhesion in composites or the interlaminar strength of the matrix [6]. Among various nanoparticles, graphene, which allows to achieve good dispersion and can be incorporated in both hydrophilic and hydrophobic polymers, is also systematically employed [7–9]. The creation of polymer nanocomposites with functionalized graphene overcomes many significant obstacles posed by filler inclusion deteriorating effects and provides superb polymer-particle interactions [10]. In this respect, the properties of the interfacial zone between the host polymer and the surface-modified nanoparticles are a tremendously important issue in successfully designing the engineering material.

3.1 Classification of Composite Materials

There is no universally accepted definition of a composite material. Polymer nanocomposites consist of at least two phases: a polymeric matrix phase and an inorganic nanofiller phase. Polymers (or macromolecules) are very large molecules made up of smaller units, called monomers or repeating units, covalently bonded together [11]. This specific molecular structure (chainlike structure) of polymeric

A. Nadtochiy et al., *Graphene-Based Polymer Nanocomposites*,
SpringerBriefs in Applied Sciences and Technology,
https://doi.org/10.1007/978-981-97-2792-6_3

materials is responsible for their intriguing mechanical properties. Polymer architecture can vary. A *linear polymer* consists of a long chain of monomers. A *branched polymer* has branches covalently attached to the main chain. *Cross-linked polymers* have monomers of one chain covalently bonded with monomers of another chain. Cross-linking results in a three-dimensional network; the whole polymer is a giant macromolecule. *Elastomers* are loosely cross-linked networks while *thermosets* are densely cross-linked networks.

Solid filling materials can be divided into four categories: polymers, metals, ceramics, and carbon, the latter can be considered separately because of its unique characteristics [12]. Therefore, the possible material combinations shown in Table 3.1 can be considered to form nanocomposites.

Composites are commonly classified by the type of material used for the matrix. Therefore, four primary classes of composites are polymer matrix composites, metal matrix composites, ceramic matrix composites, and carbon/carbon composites. Today polymer matrix composites are the most widely used class of composites. Depending upon the reinforcement, three different types of composites, i.e., particulate composites, fiber composites, and structural composites are usually considered.

Various physical properties of polymer-based composites as heterogeneous materials can be subdivided into three categories, namely *sum* properties, *combination* properties, and *product* properties [13]. Basic properties of heterogeneous materials such as electrical conductivity, dielectric constant, magnetic permeability, diffusion constant, thermal conductivity, and elastic moduli are the sum properties. For multiphase heterogeneous materials composed of n phases, the sum property parameter, K, satisfies the inequality

$$\min(K_1, K_2, ..., K_n) \leq K \leq \max(K_1, K_2, ..., K_n), \tag{3.1}$$

where $\min(\max)$ represents the minimum (maximum) value of the n property parameters, $K_i, i = 1, 2, ..., n$ is the corresponding property parameter of the i−th phase. Thus, the sum property of a piezoelectric material can be given by

$$D = eS + \varepsilon E, \tag{3.2}$$

Table 3.1 Types of composite materials [12]

			Matrix	
Reinforcement	Polymer	Metal	Ceramic	Carbon
Polymer	✓	✓	✓	✓
Metal	✓	✓	✓	✓
Ceramic	✓	✓	✓	✓
Carbon	✓	✓	✓	✓

where D is the electric displacement, e is the piezoelectric constant, S is the strain, ε is the dielectric permittivity of the material, and E is the electric field strength. The sum in (3.2) is then a measure of transforming a force field (stress) to an electric field.

Combination properties such as Poisson's ratio, electrical loss, or acoustic wave velocity are the combination of two or more basic property parameters. However, the combination property parameters do not have to satisfy the inequality given by (3.1). For example, it is known that some composites have extremely small values of Poisson's ratio, smaller than those of constituent phases.

Product properties arise from the coupling interaction between two phases with their different properties (see Table 3.2). In this Table, F represents the physical field applied to a given material, such as an electric or magnetic field. By combining different properties of two or more constituents, surprisingly large product properties can sometimes be obtained in a composite. Indeed, in some cases, product properties are found in composites that are absent in the phases making up the composites. For example, there is the magneto-electric effect in a composite material composed of one magnetostrictive and one piezoelectric phase. A magnetic field changes the shape of the magnetostrictive phase thus stressing the piezoelectric phase and generating an electric field. In this case, the coupling is mechanical. Other possible coupling effects are electric, optical, magnetic, and thermal, as seen in Table 3.2.

In Table 3.2, F_i and F_o are the input and output fields, respectively, in a two-phase heterogeneous material, and F_c is the coupling field which is the output field for one phase and the input field for another phase. As a whole, the heterogeneous material can be characterized by its $F_i - F_o$ proportionally parameter or tensor kK_1K_2 with [13]

$$\partial F_c/\partial F_i = K_1 \qquad (3.3)$$

for phase 1,

$$\partial F_o/\partial F_c = K_2 \qquad (3.4)$$

Table 3.2 Product properties of composite materials [13]

Property of phase I $F_i - F_c$	Property of phase II $F_c - F_o$	Result $F_i - F_0$
Piezomagnetism	Magnetoresistance	Piezoresistance, Phonon drag
Piezoelectricity	Electroluminescence	Piezoluminescence
Magnetostriction	Piezoelectricity	Magneto-electric effect
Magnetostriction	Piezoresistance	Magnetoresistance
Thermal expansion	Electrical conductivity	Thermistor
Hall effect	Electrical conductivity	Magnetoresistance
Piezoelectricity	Thermal expansion	Pyroelectricity
Photoconductivity	Electrostriction	Photostriction

for phase 2, and

$$\partial F_o / \partial F_i = k K_1 K_2 \tag{3.5}$$

for a composite, where k is the coupling factor ($0 \leq |k| \leq 1$) between two phases. As seen from (3.5), the product property is a more complicated linear problem.

3.2   Spatial Inhomogeneity and Free Volume of Polymers and Polymer-Based Nanocomposites

The composites are strongly heterogeneous materials with properties that vary considerably from point to point inside the material. The heterogeneous nature of composites results in complex failure mechanisms which impart toughness. Toughness is defined as the ability of a material to absorb energy before a fracture occurs. The area under a stress–strain curve for a smooth (un-notched) tension specimen loaded slowly to fracture usually characterizes toughness. The term fracture toughness can be associated with the critical stress intensity at which the final fracture occurs. In polymer composites based on thermosetting polymers, the spatial heterogeneity of the macromolecular structure occurs both at the nodular structure and the nanoscale (nanovoids or pores).

The fracture toughness depends on many factors, such as the type of loading and its preparation conditions [14]. However, the key factor is the microstructure as summarized in Fig. 3.1 [15]. The properties of nanocomposites also significantly depend on the filler shape and size.

3.2.1   Nodular Structure of Thermosets

Such morphological aspects as chain orientation and cross-linking degree (CLD) depicted in Fig. 3.1 are intrinsic peculiarities of thermosetting polymers to which epoxy resins pertain. Another distinguishing feature of thermosets is their nonhomogeneous molecular network structure, known as *nodular structure*. Nanoscale nodular morphologies in epoxy resins were heralded in the 1950s based on electron microscopy measurements [16]. More recently, comparable heterogeneous structures have been identified under ambient conditions using atomic force microscopy (AFM) at fracture interfaces [16–20]. These structures are associated with heterogeneous cross-linking and are expected to provide low energy pathways through resins, giving a structural basis for the relatively low fracture toughness of epoxy materials and rapid small molecule transport [17, 21–23]. Numerous authors have reported that these nodular structures are composed of regions of relatively higher cross-linking surrounded by an interstitial phase of relatively lower cross-link density [17, 24–26].

Fig. 3.1 Various aspects of the inner structure of polymer-based composites (reused with permissions from [15] Copyright ©2016 by the authors. Licensee MDPI, Basel, Switzerland)

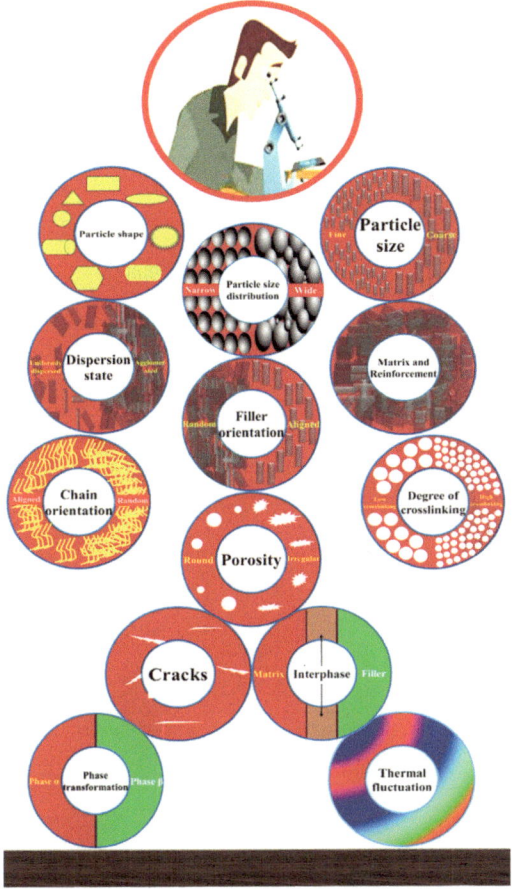

The evolution of the fracture surface morphology of the epoxy-amine system is exemplified in Fig. 3.2 [19]. It is seen that each of the three stages of network development yields a different fracture surface nodular morphology. Indeed, in the initial stage of cure, corresponding to the pregelation, active network growth, the sample shows the featureless, planar fracture surface with low levels of network development, see Fig. 3.2a. Instead, Fig. 3.2b–d taken at the network growth indicate the nodular nanostructures typical of epoxy-amine systems. The nanostructures appear at the gel point in Fig. 3.2b and become more regular near the onset of vitrification in Fig. 3.2d. After the onset, a change in the fracture surface morphology is observed in Fig. 3.2e indicative of less defined nodular nanostructures with a much rougher surface.

Fig. 3.2 AFM topographic (left) and phase (right) images of the fracture surface of the epoxy-amine system at various stages of cure. Each image represents an area of 750 nm × 750 nm. **a** 84 min (pregelation), **b** 104 min (gel point), **c** 124 min (active growth stage), **d** 184 min (active growth stage), and **e** 224 min (postvitrification) (reused with permissions from [19] Copyright ©2012 Wiley Periodicals, Inc.)

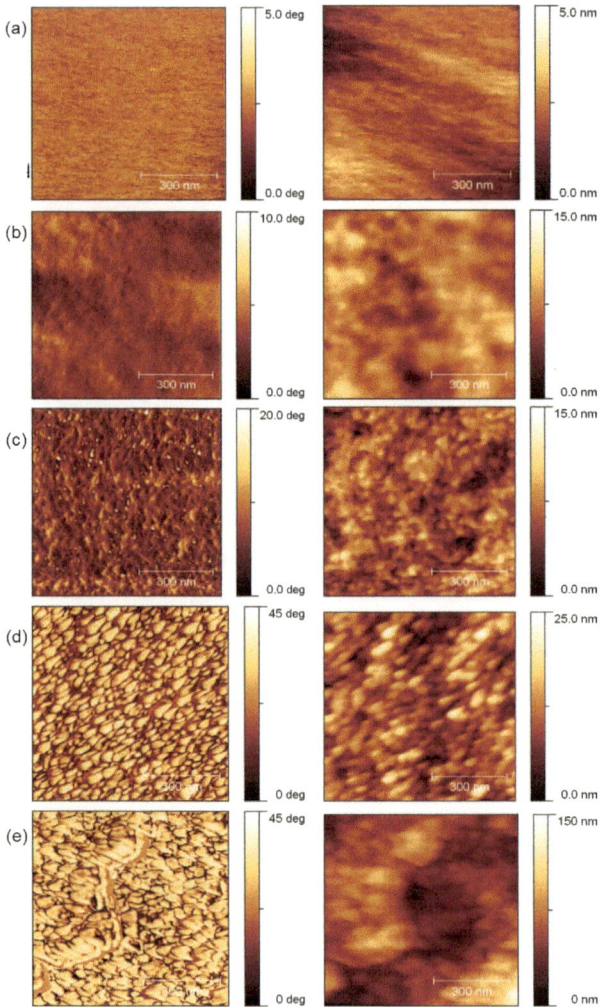

3.2.2   Free Volume Concept

In polymers, a hole or a free-volume element may be considered to be similar in size to a segment of a polymer molecule, and more than one may be required for mobility. This way, motion in polymers involves the cooperative movement of portions of a polymer chain. The free-volume concept is widely used in the theory of polymer viscoelasticity [27]. It was developed by Williams et al. [28] based on the Doolittle empirical equation [29] relating the viscosity η and the fractional free volume f_V as

$$\eta(T) = a \exp\left(\frac{b}{f_V(T)}\right), \tag{3.6}$$

where a and b are constants, T is the temperature, and f_V is given by

$$f_V(T) = \frac{V(T) - V_o(T)}{V(T)} = \frac{V_f(T)}{V(T)}, \tag{3.7}$$

where V, V_o, and V_f are a material's real, occupied, and free volumes, respectively.

There are several contributions to the specific volume (volume related to the unit of mass) of a polymer. Firstly, there is the molecular volume, a useful estimate of which is the specific Van der Waals volume, V_W, which can be calculated by group contribution methods [30]. The Van der Waals volume of a molecule is the space occupied by this molecule, which is impenetrable to other molecules with normal thermal energies. When molecules are packed in a condensed phase, there is a limit to the packing density, which can be achieved, so each molecule requires more space than its molecular volume. Typically, V_W is multiplied by a factor of 1.3, based on the packing density of a molecular crystal at 0 K. Some workers treat this as the occupied volume, so the fractional free volume can be found from (3.7) as

$$f_V(T) = \frac{V(T) - 1.3V_W(T)}{V(T)}. \tag{3.8}$$

For a variety of glassy polymers, this approach gives the values of f_V in the range from 0.11 to 0.23 [31]. The free volume may then be subdivided into interstitial-free volume, which is spread among all the molecules, and hole-free volume, which is localized but readily redistributed [32]. It is just the hole-free volume that can be included in the above theory of Williams et al.

Some scientists define the occupied volume V_o as the volume of the liquid at equilibrium at 0 K [32], which is independent of temperature. Others include the effect of thermal vibrations in V_o. A picture presented in many polymer textbooks is that the occupied volume increases approximately linearly with temperature throughout the glassy and rubbery regimes. In contrast, the free volume V_f is approximately constant in the glassy state while increasing rapidly with temperature in the rubbery state [33, 34]. Since consensus is still lacking, a semi-quantitative approach has proved very useful.

Nanovoid formation and mechanics in the context of ductile fracture toughness of thermosets have been given remarkable attention in recent years both experimentally [35–37] and theoretically [35, 38–40]. In particular, Elder et al. [40] proved that stress in the diglycidyl ether of bisphenol A (DGEBA) epoxy resin can concentrate near hollow or soft particles on the micron or sub-micron scale. This results in the dissipation of fracture energy and improved fracture toughness through shear banding between voids or plastic growth of the voids. It was also found that nanovoids in epoxy were formed in relatively non-polar regions. This in turn implies that non-polar chemistry locally enhances void formation and, conversely, that strong noncovalent interactions locally inhibit nanovoid formation [40].

It is furthermore assumed that the crack-induced stress field causes the debonding of nanoparticles in the nanocomposite's bulk, resulting in a distribution of nanovoids

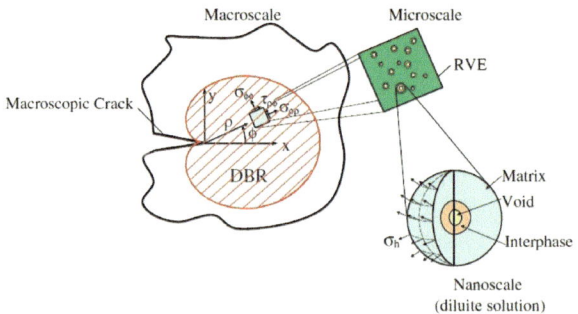

Fig. 3.3 Sketch of the multiscale (macro–micro–nanoscale) system schematics used to describe the nanovoid formation with a subsequent cracking. DBR—the debonding region, RVE—representative volume element (reused with permissions from [39] Copyright ©2012 Elsevier Ltd.)

that later undergo plastic yielding. Both the debonding of nanoparticles and the subsequent plastic yielding of nanovoids represent mechanisms acting for energy dissipation at the nanoscale. But as is generally the practice simplifying case, a debonded spherical nanoparticle of radius r_0 creates a nanovoid of the same radius, which subsequently encounters plastic deformation. This suggestion has obtained direct experimental evidence [41]. The debonding region shown in Fig. 3.3 is then thought of as the active zone of the cracking process, namely as the region around the crack tip confining all the debonded nanoparticles.

In this picture, the interphase zone surrounding the nanoparticle has properties different from those of the constituents. One could expect the priority of nanocomposites in comparison with micro-composites just in the case that the interphase presents higher material properties than the matrix. In clear contrast to micron-sized composites, in nanoscale materials, the surface effects, rather than volume, dominate due to the high surface-to-volume ratio [42]. Then the amount of interphase volume may cover a large part of the matrix and the interphase properties, which may be linked to surface functionalizers, have a profound effect on the debonding stress, especially for nanoparticles with $r_0 < 50$ nm [38]. However, a very precise knowledge of the variation of the interphase zone properties across its thickness as well as the size of the interphase zone is still obscure. It is usually assumed that a given property distribution across the interphase is some overall averaged. Consequently, the interphase is supposed to be homogeneous and isotropic [38, 43, 44].

The microstructure of the free volume and its temperature dependence in the epoxy resin DGEBA Epon 828 (with an average molecular weight of about 380 g/mol) were estimated using positron annihilation lifetime spectroscopy [45]. It was found that $f_V \approx 0.06$ at 300 K and the mean volume of nanovoids varied from 35 to 130 Å^3.

3.3 Interphase Layers in Polymer Nanocomposites

From the above brief discussion, understanding the structure and properties of the nanofiller/matrix *interface* and the resulting *interphase* region (the term *interfacial* region is used in the literature interchangeably) sketched in Fig. 3.4 are crucial for studying and controlling the properties of polymer nanocomposites [46].

As the interphase is a key performance factor in nanocomposite materials, its design and characteristics have received extensive attention. The incorporation of nanofillers into the polymer matrix was aimed mainly at improving the stress transfer between the matrix and reinforcement fillers and providing the interphase with functional properties. Enhancing the stress transfer capability is the traditional role of the interphase which is spurred by roughening the filler surface, reducing the stress transfer length thus decreasing the critical aspect ratio of the filler, and homogenizing the stress field by reducing stress concentration effects [46]. Surface roughening is beneficial both for improving the frictional component of adhesion and toughening. The crack developed at the interface or in the interphase zone is forced to propagate zigzag ledges thereby avoiding the nanofiller particles which act as local obstacles. Enhanced energy dissipation is involved in a zigzag crack path due to various nanofiller- and matrix-related failure events.

The other empirically important aspect of interphase engineering is related to functionality. It was well-recognized that some functionalities could be achieved by modifying the interphase instead of the bulk. Thus, if failure starts in the interphase, as noted above, then its detection and eventual healing should also be functionally optimized. A comprehensive survey on this field can be found elsewhere [47, 48].

Circumventing the limitations of current knowledge of interface and interfacial phenomena is critical to delivering on future promises of polymer-based nanocomposites. The literature over the past two decades treats this topic from both experimental [49–59] and theoretical [43, 60–69] perspectives.

To gain a deeper understanding of the nanoparticle-reinforced composite system, the change in the molecular structure near the particle/polymer interface is still to be considered. Molecular chain dynamics in the vicinity of nanoparticles have been reviewed by Song and Zheng [56], and the literature in this area is extensive.

Fig. 3.4 Schematics of a composite material indicating an interphase (or interfacial) region and interface formed around a filler nanoparticle embedded in a matrix

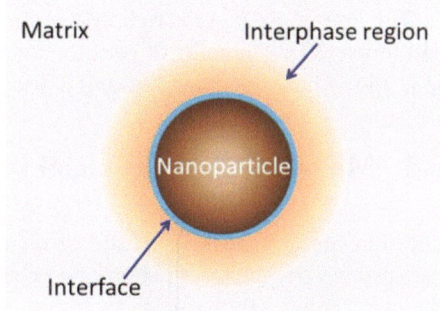

Matrix   Interphase region

Nanoparticle

Interface

The mutual arrangements of polymer chains and nanoparticles are determined by the complex contributions of several factors, including topology and concentration of nanoparticles, molecular weight and stiffness of polymers, interaction forces, etc. [65] The interaction forces include short-ranged Van der Waals attraction, long-ranged electrostatic repulsion, steric repulsion between the neighboring particles, and polymer-induced depletion attraction [70].

Due to the filler-polymer interfacial interactions, polymer segments attached at the filler surfaces [71] may form a mesophase [72] of complicated spatial chain architecture [73–75], resulting in properties in the vicinity of nanoparticles differing from the bulk [76, 77].

The filler-polymer interactions are important for nanoparticle dispersity in the matrix [78] as nanoparticles tend to precipitate locally upon relatively weak interactions and increasing polymer molecular weight [62]. The interactions are also important for heterogeneities of polymer chains located in a very close proximity to nanoparticles [79–82]. In the composites, nanoparticles immobilize a fraction of molecular chains at the surface [80] which thus exhibit a disturbed dynamics depending on filler-polymer interactions. Strong interfacial interactions increase the formation of a glassy layer whose thickness decreases with temperature [82] concomitantly reducing miscibility at a high filler volume fraction [81]. A strong polymer chain adsorption on the filler surface may promote the dispersion while the very strong bridging effect favors phase separation, so the polymer absorption can modify interparticle interactions and stabilize nanoparticles [79].

If polymer segments are not attached to the filler surface in repulsive or neutral nanoparticle-reinforced composites the phase behavior is balanced by electrostatic repulsion, polymer-induced depletion attraction, and the kinetic slowdown of diffusion-limited nanoparticle aggregation [83]. The distortions of the polymer conformation from ideality, and hence the spatial extent of the interphase, occur within the depletion region associated with nanoparticles [52]. The polymer chains close to nanoparticles tend to be elongated perpendicular to the interface [77] and interphase is not detectable essentially [84]. In contrast, faster polymer dynamics close to the interface may arise due to reduced density of the matrix [85]. It has been observed that the shape of a polymer chain is flattened tangentially to the surface at intermediate distances, and then extends radially due to entropic repulsion at closer distances. For example, in the nanocomposite of polystyrene/cross-linked PS nanoparticles, the local segment conformations can be affected in the range of a nanoparticle radius [86].

The interphase layer is rich in macromolecular tails, adsorbed segments (trains), and loops [87]. The units of parallel configuration form a loop-rich inner sublayer [88, 146] where the chains exhibit a distribution of contact lengths [89].

3.4 Models of the Interfacial Region

Different models were proposed to characterize the inclusion of nanosized filler materials in polymers. As mentioned above, they are based on the assumption that the use of filler nanoparticles yields an enlarged interfacial area between them and the matrix polymer. Then the improved properties of nanocomposites are due not only to the morphology and intrinsic properties of the filler nanoparticles alone but also to the extensive interphase created from polymer molecules of altered mobility in the vicinity of the fillers.

3.4.1 Interphase Volume Model

The *interphase volume model* [90] is based on the assumption that interphase is formed around each filler particle, independent of particle size. This interphase is supposed to have properties that are different from those of the remaining uninfluenced polymer. In general, as described above, the interfacial region in polymer nanocomposites can be characterized to extend from the surface of the filler particle, through the modification layer and interfacial polymer layers with modified chain structures, to the host matrix polymer.

Thus, the interfacial region can be considered as a shell around the particle with a certain thickness; see Fig. 3.5a. The content of the interfacial region becomes significant and even predominant in the nanocomposites when the size of fillers is reduced to the nanoscale.

To understand how much interphase is formed by using nanoscale filler materials, the volume fraction of the interfaces (f_{int}) with a finite thickness is quantitatively estimated, depending on the filler content and using the thickness of the interphase as a parameter. In the interphase volume model, the following assumptions are made [92]: (i) all filler particles are spherical and they are enclosed in interphase of constant thickness t, (ii) all filler particles have the same diameter d as shown in Fig. 3.5a, and (iii) the particles are homogeneously dispersed in the polymer matrix material. Then the volume fraction f_{int} is [91]

$$f_{int} = f_{fil}\left[\left(1 + \frac{2t}{d}\right)^3 - 1\right], \tag{3.9}$$

where f_{fil} is the filler volume fraction.

It is seen that decreasing d at a fixed interface thickness t drastically increases the volume fraction of the interfaces f_{int}. This is exemplified in Fig. 3.5b for three values of t.

One of the most fascinating features is that the nanocomposite material properties (mechanical, electrical, thermal, dielectric, etc.) in the interfacial zone differ from either the polymer matrix or the filler properties. There are generally three types of behavior at the polymer/filler interface, which are modeled in Fig. 3.5c. In this model, the properties gradually change from a value of phase A (filler) to phase B (matrix). Then the properties of the interfacial phase can normally be limited by the ones in phases A and B, as schematically illustrated by the left-hand image in Fig. 3.5c. While in the other two images in Fig. 3.5c, the properties in the interfacial zone show abnormal greater or smaller values compared to that in both phases. As the interfacial

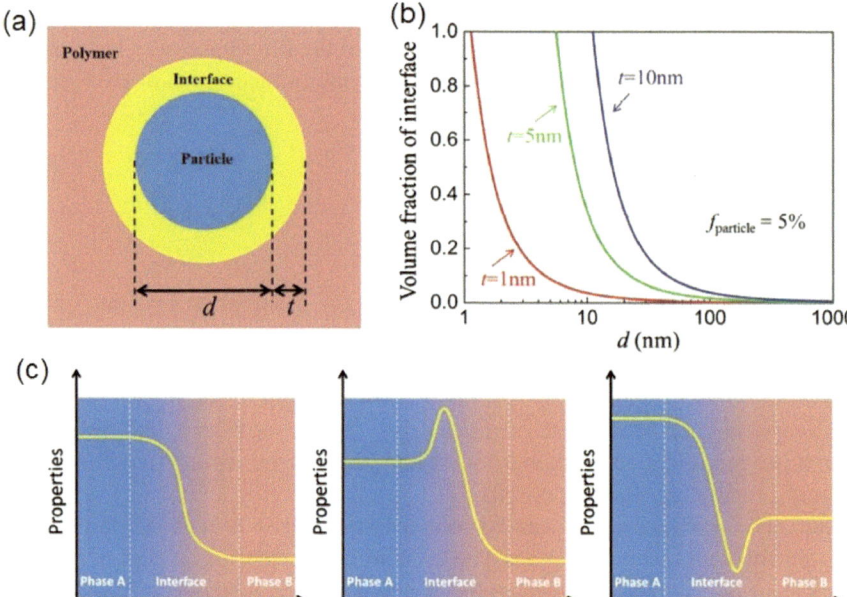

Fig. 3.5 **a** Schematics of the spherical filler/polymer interfacial structure. **b** Calculated interface volume fraction f_{int} upon filler diameter d at varying interface thickness t at the filler volume fraction fixed at 5%. **c** Illustrative examples of interfacial properties varying across the interfacial region. Here, phase A is formed by the filler and phase B by the matrix (reused with permissions from [91] Copyright ©2018 Wiley-VCH Verlag GmbH & Co. KGaA, Weinheim)

volume fraction may be quite significant, the unique properties near the interface can become predominant and thus determine the properties of the nanocomposite.

It is clear though that (3.9) is only valid if the interphases of neighboring filler particles do not overlap. Upon some simple geometry [92], this is satisfied for all values of the interparticle distance a_0 if

$$\frac{2}{\sqrt{3}}(d + 2t) \leq a_0. \tag{3.10}$$

Slightly overlapping neighboring particles shown in Fig. 3.6 are realized under certain conditions given by

$$(d + 2t) \leq a_0 < \frac{2}{\sqrt{3}}(d + 2t). \tag{3.11}$$

In this case, the interfacial overlapping parts marked by ellipses in Fig. 3.6 must be definitely subtracted from the whole interphase volume for each of the eight neighboring filler particles. Again, the geometry of the system dictates that [92]

Fig. 3.6 Geometry of interphase regions in a polymer nanocomposite filled with spherical nanoparticles: Slightly overlapping interphase zones around the nearest neighboring filler particles

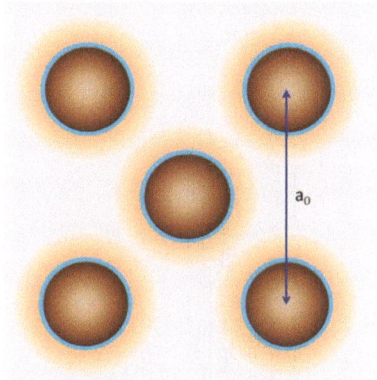

$$f_{int} = f_{fil}\left[\left(1 + \frac{2t}{d}\right)^3 - 1 - 8\left(\frac{1}{2} + \frac{t}{d} - \frac{\sqrt{3}}{4}\frac{a_0}{d}\right)^2\left(2 + \frac{4t}{d} + \frac{\sqrt{3}}{2}\frac{a_0}{d}\right)\right]. \quad (3.12)$$

Similar to the case described by (3.9), the volume fraction of the interfaces f_{int} rises linearly with the filler content f_{fil} with a relatively decreased slope.

A strong overlap of the neighboring particles takes place (Fig. 3.7) when the condition

$$\frac{2\sqrt{2}}{3}(d + 2t) \leq a_0 < (d + 2t). \quad (3.13)$$

is satisfied. In this case, the interphase of the neighboring six center particles (marked in Fig. 3.7) must be extracted additionally from the whole interphase volume to give [92]

$$f_{int} = f_{fil}\left[\left(1 + \frac{2t}{d}\right)^3 - 1 - 8\left(\frac{1}{2} + \frac{t}{d} - \frac{\sqrt{3}}{4}\frac{a_0}{d}\right)^2\left(2 + \frac{4t}{d} + \frac{\sqrt{3}}{2}\frac{a_0}{d}\right) \right.$$
$$\left. -6\left(\frac{1}{2} + \frac{t}{d} - \frac{a_0}{2d}\right)^2\left(2 + \frac{4t}{d} + \frac{a_0}{d}\right)\right]. \quad (3.14)$$

The same tendency of a decreased f_{int} at unchanged f_{fil} is observed because the fillers occupy the volume of the neighboring interphases.

Finally, if the so-called triple point is reached (Fig. 3.8) at

$$a_0 \leq \frac{2\sqrt{2}}{3}(d + 2t), \quad (3.15)$$

the whole nanocomposite material consists solely of the interphase and filler volumes. In this case, the interphase content f_{int} decreases with increasing filler content f_{fil} [92]

$$f_{int} = 1 - f_{fil}. \quad (3.16)$$

Fig. 3.7 Geometry of interphase regions in a polymer nanocomposite filled with spherical nanoparticles: Strongly overlapping interphase zones around the nearest neighboring filler particles

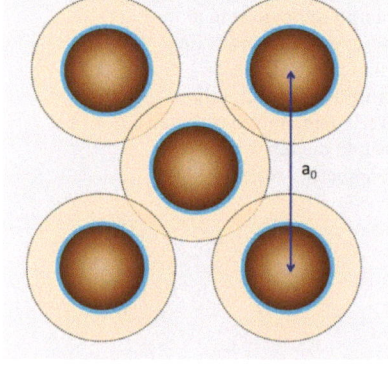

Fig. 3.8 Geometry of interphase regions in a polymer nanocomposite filled with spherical nanoparticles: Complete overlap of interphase zones around the nearest neighboring filler particles

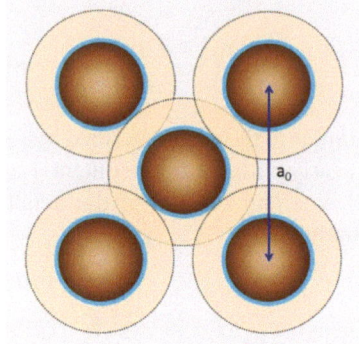

3.4.2 Multi-core Model

A *multi-core model*, or a multi-layered core model (see Fig. 3.9), was proposed to describe the interfacial electrical and dielectric behaviors in polymer nanocomposites [49] by extending the diffuse electrical double layer model into the polymer nanocomposites [93]. The interfacial zones are constructed using chemical and electrical analysis. Inorganic fillers are assumed to be spherical for simplicity. A nanoparticle of several tens of nanometers in diameter is surrounded by an interfacial layer with a thickness of the same order of magnitude as the filler diameter connecting the filler to an outside polymer matrix. An interfacial multi-layered thin shell is composed of three layers, which include a bonded layer (the first layer in Fig. 3.9), a bound layer (the second layer), and a loose layer (the third layer). Furthermore, the electric double layer is superimposed on the interfacial layer, corresponding to the Gouy–Chapman diffuse layer [94–97] with the Debye shielding length from several tens to about a hundred nanometers as also shown in Fig. 3.9. A far-field coupling of neighboring nanoparticles may appear due to their respective dipole moments.

The first layer in Fig. 3.9 with a thickness of about 1 nm corresponds to a transition layer tightly bonded to both inorganic and organic materials by coupling agents

Fig. 3.9 Schematics of the spherical filler/polymer interfacial structure in the multi-core model

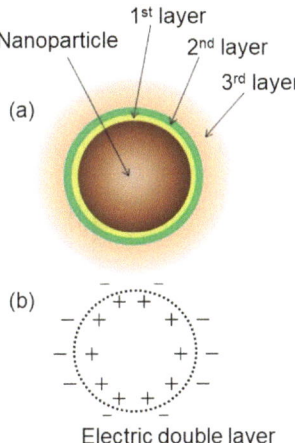

such as silane. The second layer is an interfacial region consisting of a layer of polymer chains strongly bound to and/or interacting with the first layer and the surface of the inorganic filler. Its thickness is typically in the range from 2 to 9 nm, depending on the strength of the polymer-filler interaction such that the stronger the interaction the larger the bound polymer fraction [49]. The third layer, as thick as hundreds of nanometers, is generally considered a loose layer having different chain conformation, chain mobility, and even free volume or crystallinity from the polymer matrix. The third layer is affected chemically by the second layer and electrically by the diffuse Gouy–Chapman layer.

The appearance of the Gouy–Chapman diffuse layer can be explained by a schematic picture shown in Fig. 3.10. There is an interface charge distribution between the filler particle and polymer matrix in this model. The filler which is in contact with the matrix may acquire a charge while reaching a constant chemical potential in the equilibrium thus causing an ionization of surface groups and adsorption of mobile ions from the polymer matrix. The case of a positively charged filler is exemplified in Fig. 3.10. If the polymer matrix contains mobile ions, they migrate under Coulombic forces to establish a diffuse electrical double layer around the filler particle. In this case, the double layer consists of positive and negative ions with the spatial distribution determined by the modified Poisson–Boltzmann equation solved using the Gouy–Chapman or the Debye-Hückel approximation [95, 98].

The multilayer model assuming a gradual change of the polymer dynamics in the vicinity of the filler nanoparticle interface was demonstrated in several interacting systems. The polybutadiene (PB) bound layer on carbon black (CB) nanoparticles is composed of two regions, including an inner unswollen region about 0.5 nm thick and an outer swollen region in which the polymer chains display a parabolic profile with a diffuse tail [99]. The bound layer profile in silica-filled PMMA is sensed using a small-angle neutron scattering (SANS) [100].

Fig. 3.10 Gouy–Chapman–Stern model of the electrical double layer. The diffuse electrical double layer is produced by a charged filler particle in a polymer matrix containing mobile ions. For clarity, a part of the filler surface is expanded and shown planar

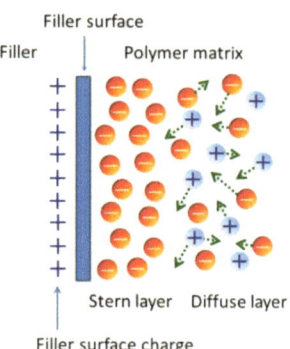

A combination of temperature-modulated differential scanning calorimetry (TMDSC), thermogravimetric analysis (TGA), and molecular modeling demonstrated different thermal activities of the adsorbed fractions of poly(vinyl acetate) on silica [101]. Compared to bulky PVA, the tightly bound segments in more direct contact with silica show a higher major thermal decomposition peak and a significantly elevated and broadened glass transition due to hydrogen-bonding interactions at the filler-polymer interface (Fig. 3.11) [102]. The loosely bound polymer further away shows only a slightly elevated transition relative to the bulk. A more mobile thermal transition for PVA at the air interface is also observed at temperatures lower than the bulk. The TMDSC results agree well with the density profiles obtained from molecular dynamics simulations (Fig. 3.12). These data present experimental and computational evidence for the existence of three distinct interfacial regions, a peak of high density, a flat region of loosely bound material, and the polymer/air interface with a size of about 1 nm.

3.4.3 Filler Shape and Size Effects

The above evidence demonstrates that the thickness and material properties of the interphase region are mainly determined by the chemical and physical characteristics and interactions of the filler and the matrix. As above, assume that the interphase layer around the filler inclusion is homogeneous and has a constant thickness h. Assume that the filler volume fraction f_{fil} is small enough to avoid overlapping interphase layers around the inclusions. Then, from the geometry of the representative volume element region, one finds the volume fraction of the interfaces [103]

$$f_{int} \geq \frac{h S_{fil}}{V_{RVE}}, \tag{3.17}$$

$$V_{RVE} = \frac{V_{fil}}{f_{fil}}, \tag{3.18}$$

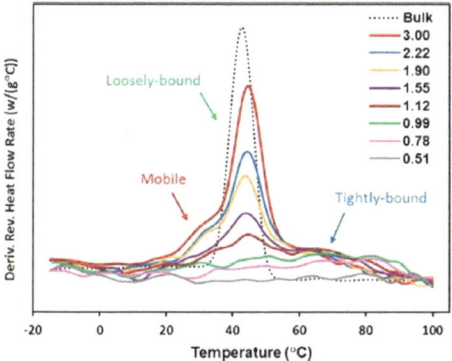

Fig. 3.11 TMDSC thermograms for bulk PVA and adsorbed PVA on silica. The thermograms are labeled with the adsorbed amounts shown in mg polymer/m^2 silica. The thermograms of the main peaks for adsorbed samples are in the same order as in the legend. The area under the tightly bound transition remained relatively constant, and the intensity of the loosely bound polymer increased with increasing adsorbed amounts of polymer (reused with permissions from [102] Copyright ©2016, American Chemical Society)

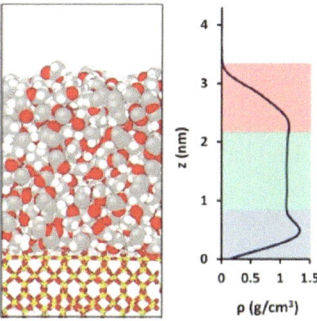

Fig. 3.12 Simulated snapshot side view of adsorbed PVA on silica (left) and the density profile of the polymer as a function of the distance from the surface (right). Blue (upper), green (middle), and red (lower) areas in the density profile highlight the tightly bound, loosely bound, and mobile regions of PVA, respectively (reused with permissions from [102] Copyright ©2016, American Chemical Society)

where S_{fil} and V_{fil} are the surface area and the volume of the filler particle, respectively, and V_{RVE} is the volume of the RVE. From (3.17) and (3.18), one gets

$$f_{int} \geq f_{fil} h \frac{S_{fil}}{V_{fil}}, \tag{3.19}$$

It is clear from (3.19) that the volume fraction of the interphase, $f_{int} = f_{int}(f_{fil}, h, S_{fil}/V_{fil})$, is a function of the volume fraction of the filler particle, the thickness of the interphase and the surface-to-volume-ratio of the filler. The thinner

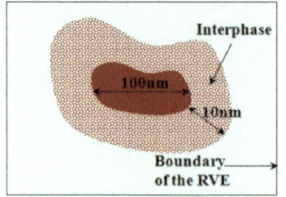 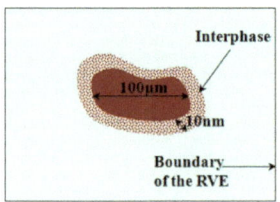

Fig. 3.13 Schematic comparison of the filler/matrix length scales for nm-(left) and μm-(right) sized fillers. Dimensions of the interphase and RVE are not to scale (reused with permissions from [103] Copyright ©2010 Elsevier Ltd.)

the interphase, the closer the value of the volume fraction of the interphase to the lower value predicted by (3.19). Since S_{fil}/V_{fil} is size-dependent, f_{int} is size-independent even if the thickness h is size-independent. This behavior is schematically illustrated in Fig. 3.13.

The involvement of the filler shape and size is a synergy of several factors, such as the volume fraction of the filler, the thickness of the interphase, and the size and morphology of the filler particle. The effects of a filler size are mainly due to the size-dependent surface-to-volume ratio of the filler. To study the effects related to the morphology of the inclusion, the shape variations of the surface-to-volume ratio should be taken into account [103]. Besides spherical filler nanoparticles, two geometrical shapes, cylinders and disks, have most frequently been considered in any of the investigations; see Fig. 3.14. Nanoplatelets are usually considered to be isotropic solid round disks with a radius R_d, while nanotubes or fibers are approximated by the cylinders having a length L. Furthermore, cylinders with different length aspect ratios can represent the fiber-like particle and the platelet illustrated in Fig. 3.14a, b. The elastic modulus and Poisson ratio are typically chosen to reflect moduli appropriate for individual single-walled nanotubes or individual graphene sheets.

Different filler types with the same volume have a different surface area. Thus, for the spherical particles with a radius R_s shown in Fig. 3.14c

$$V_{fil}^s = \frac{4}{3}\pi R_s^3, \ S_{fil}^s = 4\pi R_s^2. \tag{3.20}$$

For the cylindrical particles given by Fig. 3.14a, b

$$V_{fil}^c = \pi R_d^2 L, \ S_{fil}^c = 2\pi R_d^2 + 2\pi R_d L. \tag{3.21}$$

Then the length aspect ratios (the dominant dimension over the minor dimension of the filler) of the fiber-like particle n_f and that of the platelet n_p are defined as [103]

$$n_f = \frac{L}{2R_d}, \tag{3.22}$$

Fig. 3.14 Schematics of **a**
cylinder, **b** disk, and **c** sphere
used to approximate
nanotube, nanoplatelet, and
spherical filler particles,
respectively

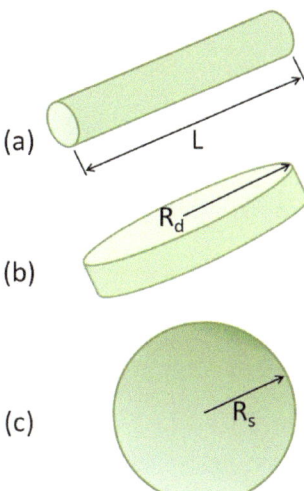

(a)

(b)

(c)

$$n_p = \frac{2R_d}{L}. \tag{3.23}$$

Defining the characteristic length $\mathcal{L}$ of particles as

$$\mathcal{L} = \sqrt[3]{V_{fil}}, \tag{3.24}$$

the surface-to-volume-ratios of spherical, fiber-like, and platelet-like particles, respectively, can be given as [103]

$$\left(\frac{S_{fil}}{V_{fil}}\right)_s = \sqrt[3]{36\pi}\ \mathcal{L}^{-1}, \tag{3.25}$$

$$\left(\frac{S_{fil}}{V_{fil}}\right)_f = \sqrt[3]{2\pi n_f}\ \frac{2n_f + 1}{n_f}\ \mathcal{L}^{-1}, \tag{3.26}$$

$$\left(\frac{S_{fil}}{V_{fil}}\right)_p = \sqrt[3]{2\pi n_p}\ \frac{2n_p + 1}{n_p}\ \mathcal{L}^{-1}. \tag{3.27}$$

It is seen in Fig. 3.15a that, for the same characteristic length $\mathcal{L}$, the platelet has a much larger surface-to-volume ratio than the fiber-like and spherical particles. It is also seen in Fig. 3.15b that the size effect becomes more significant with increasing the length aspect ratio n_f of the platelets. As illustrated in Fig. 3.15, the surface-to-volume ratios of the fillers show strong size effects when their characteristic lengths are smaller than 10–20 nm. A rough estimate shows that the value $\mathcal{L} = 20$ nm is obtained with the 1-nm thick platelet having the length aspect ratio $n_p = 100$.

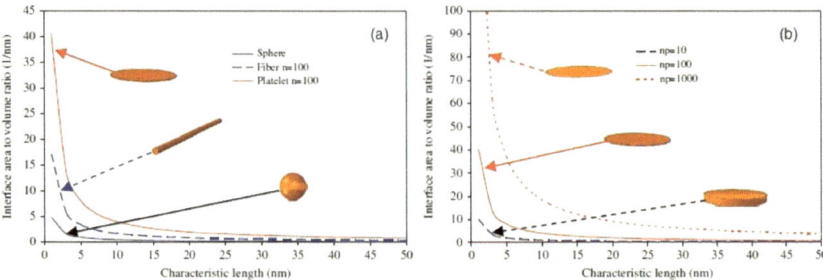

Fig. 3.15 Values of S_{fil}/V_{fil} in (3.25)–(3.27) upon varying characteristic length $\mathfrak{L}$ **a** for different particle shapes and **b** for platelets with various aspect ratios n_p (reused with permissions from [103] Copyright ©2010 Elsevier Ltd.)

As a final remark, the volume of non-overlapping interphase layers for disk-shaped nanoplatelets is [104]

$$V_{int} = \left[\pi(R_d + t)^2 - \pi R_d^2\right]h + 2\pi R_d^2 t + \frac{4}{3}\pi t^3 + \pi^2 R_d t^2. \tag{3.28}$$

Then

$$\frac{f_{int}}{f_{fil}} = \frac{t}{R_d}\left[2 + \frac{t}{R_d} + 2\frac{R_d}{h} + \frac{t}{h}\left(\pi + \frac{4}{3}\frac{t}{R_d}\right)\right]. \tag{3.29}$$

3.5   Mechanical and Electrical Percolation in Nanocomposites

Improved mechanical properties are observed in polymer nanocomposites at extremely low filler volume fractions [105]. As discussed above, these effects can be due in part to the significant perturbation of the matrix material by filler particles and the formation of the interfacial region in nanocomposites. This may increase the stiffness of the polymer in a thin-layer surrounding the reinforcing fillers, effectively creating a composite interphase. These local effects are pronounced in all composite materials, but because of the high surface-to-volume ratio at the nanoscale, the interfacial volume fraction in nanocomposites can be greater than that of the particles, representing a significant, stiffer than the matrix, composite phase [106]. In addition to contributing to the overall effective stiffness of the composite, it has been proposed that these interfacial regions contribute to the formation of percolated microstructures by forming connections between particles and interface, a pseudo-percolation, or by percolating themselves [106–109].

Dramatic changes in the physical properties of the composites, including their electrical or thermal conductivity, when the filler content is near the threshold are schematically illustrated in Fig. 3.16 [110].

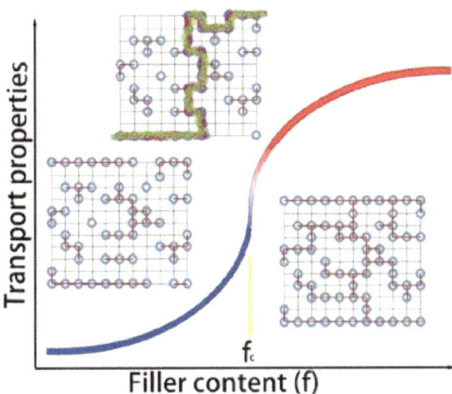

Percolation theory describes the non-linear scaling of transport properties with loading level in statistically filled systems due to the random nature of particle–particle connections. A universal scaling relation for electrical conductivity σ according to percolation theory relates the conductivity of the filler particles σ_0, the volume fraction f_{fil}, the volume fraction of filler at percolation threshold f_c, and a scaling exponent μ and μ' in the conducting and insulating regions, respectively, by [110, 111]

$$\sigma = \sigma_0(f_{fil} - f_c)^\mu \tag{3.30}$$

for $f_{fil} > f_c$ and

$$\sigma = \sigma_0(f_c - f_{fil})^{-\mu'} \tag{3.31}$$

for $f_{fil} < f_c$.

If the filler is distributed randomly in the system the scaling exponent value is dictated only by its dimensionality, so that $\mu \approx 1$–1.3 in 2D systems like thin films, and $\mu \approx 1.6$–2 in three-dimensional systems like bulk composites [111–113]. Experimental data are indeed frequently consistent with these values [114], whereas observed deviations of the measured μ from the universal values are ascribed merely to varying spatial distributions of the conducting particles [111]. For example, broadening of the interparticle contact resistance distribution increases μ, so the exponents about twice as much as the expected value [115] or even as high as ≈ 12 [116] have been reported. In contrast, when the spatial distribution of particles in the composite shows strong local correlations, a reduction in μ below the universal value may be obtained [113].

The computed and measured electrical conductivities of polymerized cyclic butylene terephthalate (pCBT)-based composites containing nanofillers of graphene nanoplatelets (GNPs) are compared in Fig. 3.17 [117]. It is seen that the bulk electrical conductivity of composites containing GNPs with $n_p = 100$ is smaller than that in composites with $n_p = 1000$. Moreover, the percolation threshold f_c is about 4 wt.% for composites with $n_p = 100$ and 1 wt.% in composites with $n_p = 1000$.

Fig. 3.17 Electrical conductivities of GNPs/pCBT composites with two aspect ratio values ($n_p = 100$ and 1000) of GNPs, experimentally measured and obtained from the micromechanics modeling [117] (reused with permissions from [117] Copyright ©2014 Elsevier Ltd.)

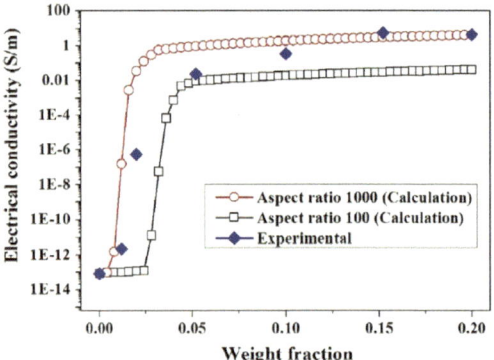

As can be seen, the micromechanics modeling matches the observed experimental data rather well.

Another example is shown in Fig. 3.18 [118] illustrating the dependence of the electrical conductivity upon the volume fraction of the filler multi-walled carbon nanotube (MWCNT) and graphene nanosheet (GNS) particles in the MWCNT/HDPE and GNS/HDPE composites.

Percolation behavior is observed in both the composites. The percolation threshold f_c of MWCNT/HDPE composite is about 0.15 vol.% (0.32 wt.%). This value is considerably smaller than those obtained with a random carbon nanotube distribution. Thus, the percolation threshold was found to be 2 wt.% for the MWCNT/HDPE composites prepared by solution-precipitation [119], 4 wt.% for those obtained by melt processing [120], and 3.1 wt.% for the composite processed by solution casting-drawing using mixed solvents [121]. The percolation threshold for GNS/HDPE composite with a segregated network structure is 1 vol.%, greater than that for MWCNT/HDPE composites.

The smaller percolation threshold for MWCNT/HDPE in Fig. 3.18 was explained by the fact that a distribution of MWCNTs can efficiently increase their

Fig. 3.18 An example of the percolation threshold in high-density polyethylene (HDPE) composites formed with MWCNTs (1) and (GNSs) (2). Inset shows a log-log plot of the conductivity as a function of $f_{fil} - f_c$ (reused with permissions from [118] Copyright ©2010 Elsevier Ltd.)

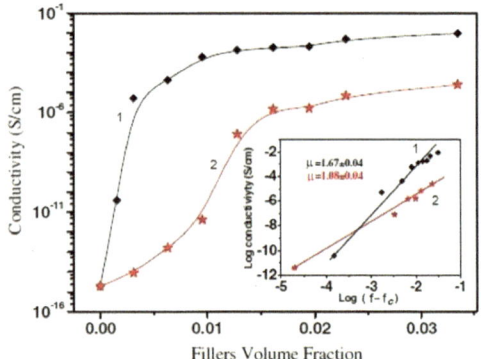

utilization ratio in forming a conducting network at lower filler loading [118]. Moreover, MWCNT/HDPE composites have two orders of magnitude higher conductivity than GNS/HDPE composites with the same f_{fil}, implying that MWCNTs are a more efficient conductive filler.

3.5.1 Modeling Approach

The models used to describe percolation invariably rely upon prior knowledge of a theoretical percolation threshold. Models based on percolation thresholds are well developed in the study of electrically conducting composites. Meanwhile, due to the resemblance of the mechanical and electrical percolation curves, the electrical models have been frequently used to describe mechanical effects by replacing conductivity terms with stiffness terms. However, there is a difference in mechanical and electrical behaviors. Below the threshold percolation volume fraction, the composite typically has low conductivity, while above the threshold it is greatly enhanced. In turn, mechanical percolation frequently has more intermediate stages. A connected filler network would enhance mechanical properties, but composite properties are continuously affected by the volume fraction of the filler.

To model mechanical percolation, McLachlan et al. [122] exploited the Generalized Effective Media model, which combined a mean-field model at low volume fractions and percolation theory above the percolation threshold. Both electrical and mechanical percolation have been successfully described by this model [123]. Another model that has been used is the series-parallel model [124]. It includes an intermediate parameter of the material volume fraction active in the transfer of forces. The required percolation threshold as an input parameter poses a limitation of these models. Seidel and Lagoudas [125] studied the influence of an interface region and the effects of clustering using the concentric cylinder micromechanics model, but not in the context of percolation thresholds. A hybrid numerical analytic model was used by Liu and Brinson [126] to investigate polymer nanocomposites with complex microstructural configurations.

Fralick et al. [127] studied percolation effects by simulating populations of random microstructures for a nanocomposite composed of fillers, matrix, and interface at discrete volume fractions. This model defines the percolation threshold and predicts the probabilistic distribution of the composite properties varied with the filler volume fraction [128].

A significant increase in stiffness was observed in the composite system when filler particles and interface regions formed connected pathways known as a pseudo-percolation [108]. The distribution of the resulting properties was likely due to a combination of mechanisms dominated by microstructure effects, spatial arrangements, percolation, and/or pseudo-percolation. To clarify the relative contributions of these effects, the results of Fralick et al. [127] discussed in the framework of the grid-based computational Generalized Method of Cells [129, 130] were compared to the predictions of the Mori–Tanaka micromechanics model [131, 132].

It is known that the shape of the graphene is not strictly a disc shown in Fig. 3.14b but it is something between a disc and a cuboid or on the other words a fileted cuboid with rounded corners. Graphene nanofillers with fileted, cuboid, and discs shapes randomly distributed in the matrix are displayed in Fig. 3.19.

Mazaheri et al. [133] developed an analytical model to describe the effective electrical conductivity and percolation behavior of exfoliated polymer-graphene nanocomposites. Graphene layers are considered as exfoliated fileted cuboids randomly distributed in the matrix. The effect of filler shape including the radius of graphene nanofillers in the shape of fileted cuboids along with tunneling and interphase effects on the percolation behavior and effective electrical conductivity of the nanocomposites have been taken into account. The model predicts that the thickness and electrical conductivity of the interphase layer play a significant role in the effective electrical conductivity of the polymer-graphene nanocomposites (see Fig. 3.20).

It was also found that the percolation threshold decreases with increasing the interphase thickness. On the other hand, the percolation threshold strongly depends on the aspect ratio n_f (Fig. 3.21). It can be found that the percolation threshold is inversely proportional to the aspect ratio of the fillers, $f_c \approx n_f^{-1}$.

The percolation threshold and ultimate direct current conductivity data obtained in different graphene-polymer nanocomposites are collected in Table 3.3.

Fig. 3.19   a Composite with randomly distributed graphene fillers with fileted shape by an interphase layer (with thickness t), **b** cuboid shape graphene nanofiller, and **c** disc shape graphene nanofiller (reused with permissions from [133] Copyright ©2020 Elsevier Ltd.)

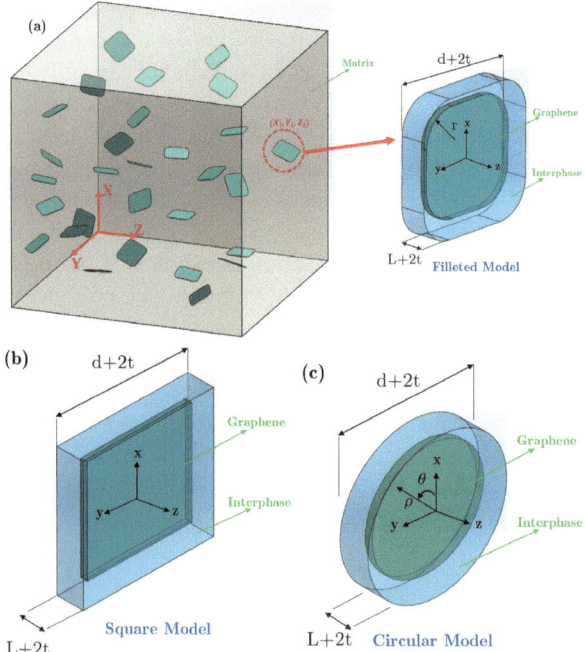

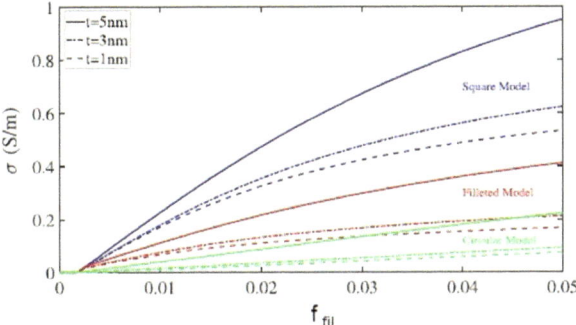

Fig. 3.20 Computed dependence of the effective electrical conductivity of polymer-graphene nanocomposites on the volume fraction of the filler particle at different interphase thicknesses (reused with permissions from [133] Copyright ©2020 Elsevier Ltd.)

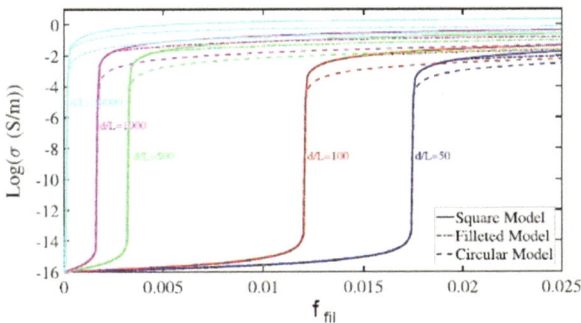

Fig. 3.21 The dependence of the effective electrical conductivity of PGNs on volume fraction with different aspect ratios ($n_f = d/L$ in this plot) (reused with permissions from [133] Copyright ©2020 Elsevier Ltd.)

Table 3.3 A summary of conductive properties of graphene-based polymer nanocomposites [134]

Polymer	Filler	Preparation method	f_c	σ (S m^{-1})	References
ABS	GO	Coagulation blending	0.13 vol.%	0.1	[135]
PA6	TRGO		7.5 wt.%	7.1×10^{-3}	[136]
PA6	CB	Melt compounding	7.5 wt.%		[136]
PA6	rGO	In situ poly-merization	0.41 vol.%	0.028	[137]
PA11	GNP	Masterbatch extrusion	0.25 vol.%	5.2×10^{-6}	[138]
PA12	TRGO	Melt compounding	1–2.5 wt.%	5.2×10^{-6}	[139]

Table 3.3 (continued)

Polymer	Filler	Preparation method	f_c	σ (S m^{-1})	References
PA12	N-TRGO	Melt compounding	1 wt.%	10^{-4}	[139]
PC	TRGO	Melt compounding	2.5 wt.%	0.1	[136]
PC	CB	Melt compounding	2.5 wt.%	9.1	[136]
PC	rGO	Solution blending/In situ thermal reduction	0.21	0.1	[140]
PC	fGNP	Emulsion mixing	0.14 vol.%	51	[141]
POSS-PCL	rGO	Solution blending	2.5 vol.%	0.1	[142]
PE	fGNP	Melt mixing	0.83 vol.%	0.01	[143]
LLDPE	TRGO	Melt mixing	0.5 vol.%	10^{-4}	[144]
LLDPE	TRGO	Melt compounding	0.5–0.9 vol.%	10^{-4}	[144]
HDPE	fGNP	Solution blending	0.37–0.74 vol.%	27	[145]
UHMWPE	kGO	Solution blending and hydrazine reduction	0.028 vol.%	5	[146]
PLA	GNP	Solution blending	0.004 vol.%	0.1	[147]
PLA	TRGO	In situ polymerization	0.5–0.75 vol.%	0.01	[148]
PMMA	rGO	Solution blending	0.25 vol.%	0.01	[149]
PMMA	fGNP	Solution blending	0.3	10^{-3}	[150]
PMMA	fGNP	Solution blending	0.8 vol.%	20	[151]
iPP	TRGO	Melt compounding	5 wt.%	5.3×10^{-2}	[136]
iPP	CB	Melt compounding	5 wt.%	3.3	[136]
iPP	CNT	Melt compounding	5 wt.%	3.1	[136]
TPU/PP	frGO	Solution-flocculation and melt mixing	0.054 vol.%	10^{-6}	[152]

Table 3.3 (continued)

Polymer	Filler	Preparation method	f_c	σ (S m^{-1})	References
PP	TRGO	Melt compounding	<5 wt.%	10^{-4}	[153]
PP	MLG	Melt compounding	5 wt.%	3×10^{-3}	[153]
PP	CB	Melt compounding	7.5 wt.%	3×10^{-5}	[153]
PP	CNT	Melt compounding	7.5 wt.%	4×10^{-6}	[153]
PS	rGO	LbL assembly	0.2 vol.%	0.05	[154]
PS	GNP	Solution blending	0.1 vol.%	13.8	[155]
PS	GNP	Solution blending	0.33 vol.%	3.5	[156]
PS	GNP	Electrostatic self-assembly	0.09 vol.%	25	[157]
sPS	GNP	Solution blending	0.46 vol.%	470	[158]
PS	GNP	Electrostatic assembly	0.054 vol.%	46	[159]
PU	fTRGO	In situ poly-merization	0.5–2 wt.%	1.4×10^{-7}	[160]
PU	CB	In situ poly-merization	2 wt.%	1.3×10^{-7}	[160]
PU	CNT	In situ poly-merization	2 wt.%	1.9×10^{-7}	[160]
PU	rGO	Solution blending	0.078 vol.%		[161]
PVC	GNP	Solution blending	0.1 vol.%	5.8	[162]
SAN	TRGO	Melt compounding	4 wt.%	0.123	[136]
SAN	CB	Melt compounding	4 wt.%	9	[136]
SAN	CNT	Melt compounding	12 wt.%	7.4×10^{-4}	[136]
Epoxy	fGNP	Solution blending	1.3 vol.%	10^{-6}	[163]
Epoxy	frGO	Solution blending	0.71 vol.%	10^{-6}	[164]
Epoxy	fGNP	Sonication/ Calendaring	4 vol.%	10^{-4}	[165]
Epoxy	GNP	Solution blending	0.5–1 vol.%	10^{-2}	[166]
Epoxy	GNP	LbL	0.6 vol.%	10^{-4}	[167]

Table 3.3 (continued)

Polymer	Filler	Preparation method	f_c	σ (S m^{-1})	References
Epoxy	TRGO	Solvent free mixing method	1 wt.%	2×10^{-6}	[168]
Epoxy	N-TRGO	Solvent free mixing method	1 wt.%	10^{-5}	[168]
Epoxy	DMG	Solvent free mixing method	5 wt.%	8×10^{-5}	[168]
Epoxy	GNP	Solution blending	0.52 vol.%	0.05	[169]
Epoxy	fGNP	Solution blending	0.16 vol.%	10	[170]
Epoxy	GNP	Three roll mill	0.22 vol.%	10^{-5}	[171]
Epoxy	fGO	Solution blending	0.78 vol.%	1	[172]
PI	fGNP	In situ poly-merization	0.5 vol.%	0.1	[173]
PI	fGO	In situ poly-merization	0.25 vol.%	0.092	[174]
NR	GNP	Latex self-assembly	0.62 vol.%	0.03	[175]
NR	CrGO	Coagulation method	3 wt.%	10^{-4}	[176]
NR	rGO	Solution blending	0.21 vol.%	0.23	[177]

References

1. T. Keller, Recent all-composite and hybrid fiber reinforced polymer bridges and buildings. Prog. Struct. Eng. Mater. **3**, 132–140 (2001)
2. B. Ramezanzadeh, Z. Haeri, M. Ramezanzadeh, A facile route of making silica nanoparticles-covered graphene oxide nanohybrids (SiO$_2$-GO); fabrication of SiO$_2$-GO/epoxy composite coating with superior barrier and corrosion protection performance. Chem. Eng. J. **303**, 511–528 (2016)
3. C. Leopold, T. Augustin, T. Schwebler, J. Lehmann, W.V. Liebig, B. Fiedler, Influence of carbon nanoparticle modification on the mechanical and electrical properties of epoxy in small volumes. J. Colloid Interface Sci. **506**, 620–632 (2017)
4. W. Han, J. Zhou, Q. Shi, Research progress on enhancement mechanism and mechanical properties of FRP composites reinforced with graphene and carbon nanotubes. Alex. Eng. J. **64**, 541–579 (2023)
5. A.A. Ropalekar, R.R. Ghadge, N. Anekar, A review on functionalization methods of graphene oxide for enhancement in mechanical properties of epoxy composites. Mater. Today: Proc. In Press, https://doi.org/10.1016/j.matpr.2023.09.098

6. T. Bao, Z. Wang, Y. Zhao, Y. Wang, X. Yi, Improving tribological performance of epoxy composite by reinforcing with polyetheramine-functionalized graphene oxide. J. Mater. Res. Technol. **12**, 1516–1529 (2021)
7. W. Zheng, S.-C. Wong, Electrical conductivity and dielectric properties of PMMA/expanded graphite composites. Compos. Sci. Technol. **63**, 225–235 (2003)
8. J. Fu, M. Zhang, L. Liu, L. Xiao, M. Li, Y. Ao, Layer-by-Layer electrostatic self-assembly silica/graphene oxide onto carbon fiber surface for enhance interfacial strength of epoxy composites. Mater. Lett. **236**, 69–72 (2019)
9. Y. Liu, A. Zou, G.-D. Wang, C. Han, E. Blackie, Enhancing interlaminar fracture toughness of CFRP laminates with hybrid carbon nanotube/graphene oxide fillers. Diam. Relat. Mater. **128**, 109285 (2022)
10. T. Ramanathan, A.A. Abdala, S. Stankovich, D.A. Dikin, M. Herrera-Alonso, R.D. Piner, D.H. Adamson, H.C. Schniepp, X. Chen, R.S. Ruoff, S.T. Nguyen, I.A. Aksay, R.K. Prud'Homme, L.C. Brinson, Functionalized graphene sheets for polymer nanocomposites. Nature Nanotech. **3**, 327–331 (2008)
11. W.D. Callister, *Materials Science and Engineering. An Introduction*, 7th edn. (Wiley, New York, 2006)
12. C. Zweben, Composite materials, in *Handbook of Materials Selection*, ed. by K. Myer (Wiley, New York, 2002), pp.357–399
13. C.-W. Nan, Physics of inhomogeneous inorganic materials. Prog. Mater. Sci. **37**, 1–116 (1993)
14. M.S. Saharudin, R. Atif, I. Shyha, F. Inam, The degradation of mechanical properties in polymer nano-composites exposed to liquid media - a review. RSC Adv. **6**, 1076–1089 (2016)
15. R. Atif, I. Shyha, F. Inam, Mechanical, thermal, and electrical properties of graphene-epoxy nanocomposites - a review. Polymers **8**, 281 (2016)
16. S. Morsch, Y. Liu, P. Greensmith, S.B. Lyon, S.R. Gibbon, Molecularly controlled epoxy network nanostructures. Polymer **108**, 146–153 (2016)
17. C.M. Sahagun, S.E. Morgan, Thermal control of nanostructure and molecular network development in epoxy-amine thermosets. ACS Appl. Mater. Interfaces **4**, 564–572 (2012)
18. M.R. Vanlandingham, R.F. Eduljee, J.W. Gillespie, Relationships between stoichiometry, microstructure, and properties for amine-cured epoxies. J. Appl. Polym. Sci. **71**, 699–712 (1999)
19. C.M. Sahagun, K.M. Knauer, S.E. Morgan, Molecular network development and evolution of nanoscale morphology in an epoxy-amine thermoset polymer. J. Appl. Polym. Sci. **126**, 1394–1405 (2012)
20. S. Morsch, Y. Liu, S.B. Lyon, S.R. Gibbon, Insights into epoxy network nanostructural heterogeneity using AFM-IR. ACS Appl. Mater. Interfaces **8**, 959–966 (2016)
21. P.J. Aspbury, W.C. Wake, The supermolecular structures found in cured epoxy resins. Br. Polym. J. **11**, 17–27 (1979)
22. K. Frank, J. Wiggins, Effect of stoichiometry and cure prescription on fluid ingress in epoxy networks. J. Appl. Polym. Sci. **130**, 264–276 (2013)
23. G. Bouvet, N. Dang, S. Cohendoz, X. Feugas, S. Mallarino, S. Touzain, Impact of polar groups concentration and free volume on water sorption in model epoxy free films and coatings. Prog. Org. Coat. **96**, 32–41 (2016)
24. U.T. Kreibich, R. Schmid, Inhomogeneities in epoxy resin networks. J. Polym. Sci. Polym. Symp. **53**, 177–185 (1975)
25. H. Kishi, T. Naitou, S. Matsuda, A. Murakami, Y. Muraji, Y. Nakagawa, Mechanical properties and inhomogeneous nanostructures of dicyandiamide-cured epoxy resins. J. Polym. Sci. B Polym. Phys. **45**, 1425–1434 (2007)
26. J. Feng, K.R. Berger, E.P. Douglas, Water vapor transport in liquid crystalline and non-liquid crystalline epoxies. J. Mater. Sci. **39**, 3413–3423 (2004)
27. Y.-H. Lin, *Polymer Viscoelasticity: Basics, Molecular Theories, Experiments and Simulations*, 2nd edn. (World Scientific, Singapore, 2010)
28. M.L. Williams, R.F. Landel, J.D. Ferry, The temperature dependence of relaxation mechanisms in amorphous polymers and other glass-forming liquids. J. Am. Chem. Soc. **77**, 3701–3707 (1955)

29. A.K. Doolittle, Studies in Newtonian flow. II. The dependence of the viscosity of liquids on free-space. J. Appl. Phys. **22**, 1471–1475 (1951)
30. D.W. van Krevelen, Properties of Polymers, ed. by K. te Nijenhuis, 4th edn. (Elsevier, Amsterdam, 2009)
31. A. Thran, G. Kroll, F. Faupel, Correlation between fractional free volume and diffusivity of gas molecules in glassy polymers. J. Polym. Sci. B Polym. Phys. **37**, 3344–3358 (1999)
32. H.F. Mark, N.M. Bikales, C.G. Overberger, G. Menges (eds.), *Encyclopedia of Polymer Science and Engineering*, vol. 5, 2nd edn. (Wiley, New York, 1986)
33. L.H. Sperling, *Introduction to Physical Polymer Science*, 4th edn. (Wiley, Hoboken, NJ, 2006)
34. U.W. Gedde, *Polymer Physics* (Springer, 1999)
35. A. Kearney, C.S. Litteken, C. Mohler, M.E. Mills, R.H. Dauskardt, Pore size scaling for enhanced fracture resistance of nanoporous polymer thin films. Acta Mater. **56**, 5946–5953 (2008)
36. D.B. Knorr, K.A. Masser, R.M. Elder, T.W. Sirk, M.D. Hindenlang, J. Yu, A.D. Richardson, S.E. Boyd, W.A. Spurgeon, J.L. Lenhart, Overcoming the structural versus energy dissipation trade-off in highly crosslinked polymer networks: ultrahigh strain rate response in polydicyclopentadiene. Compos. Sci. Technol. **114**, 17–25 (2015)
37. E.D. Bain, D.B. Knorr, A.D. Richardson, K.A. Masser, J. Yu, J.L. Lenhart, Failure processes governing high-rate impact resistance of epoxy resins filled with core-shell rubber nanoparticles. J. Mater. Sci. **51**, 2347–2370 (2015)
38. M. Zappalorto, M. Salviato, M. Quaresimin, Influence of the interphase zone on the nanoparticle debonding stress. Compos. Sci. Technol. **72**, 49–55 (2011)
39. M. Zappalorto, M. Salviato, M. Quaresimin, A multiscale model to describe nanocomposite fracture toughness enhancement by the plastic yielding of nanovoids. Compos. Sci. Technol. **72**, 1683–1691 (2012)
40. R.M. Elder, D.B. Knorr, J. Andzelm, J.L. Lenhart, T.W. Sirk, Nanovoid formation and mechanics: a comparison of poly(dicyclopentadiene) and epoxy networks from molecular dynamics simulations. Soft Matter **12**, 4418–4434 (2016)
41. T.H. Hsieh, A.J. Kinloch, K. Masania, A.C. Taylor, S. Sprenger, The mechanisms and mechanics of the toughening of epoxy polymers modified with silica nanoparticles. Polymer **51**, 6284–6294 (2010)
42. P.M. Ajayan, L.S. Schadler, P.V. Braun, *Nanocomposite Science and Technology*, 4th edn. (Wiley, Weinheim, 2003)
43. G.M. Odegard, T.C. Clancy, T.S. Gates, Modeling of mechanical properties of nanoparticle/polymer composites. Polymer **46**, 553–562 (2005)
44. S. Yu, S. Yang, M. Cho, Multi-scale modeling of cross-linked epoxy nanocomposites. Polymer **50**, 945–952 (2009)
45. G. Dlubek, E.M. Hassan, R. Krause-Rehberg, J. Pionteck, Free volume of an epoxy resin and its relation to structural relaxation: evidence from positron lifetime and pressure-volume-temperature experiments. Phys. Rev. E **73**, 031803 (2006)
46. A.N. Netravali, K.L. Mittal (eds.), *Interface/Interphase in Polymer Nanocomposites* (Wiley, Hoboken, NJ, 2017)
47. M. Sharma, S.L. Gao, E. Mader, H. Sharma, L.Y. Wei, J. Bijwe, Carbon fiber surfaces and composite interphases. Composites Sci. Technol. **102**, 35–50 (2014)
48. J. Karger-Kocsis, H. Mahmood, A. Pegoretti, Recent advances in fiber/matrix interphase engineering for polymer composites. Prog. Mater. Sci. **73**, 1–43 (2015)
49. T. Tanaka, M. Kozako, N. Fuse, Y. Ohki, Proposal of a multi-core model for polymer nanocomposite dielectrics. IEEE Trans. Dielectr. Electr. Insul. **12**, 669–681 (2005)
50. J.A. Khan, S.E. Harton, P. Akcora, B.C. Benicewicz, S.K. Kumar, Polymer crystallization in nanocomposites: spatial reorganization of nanoparticles. Macromolecules **42**, 5741–5744 (2009)
51. P.A. Klonos, A. Panagopoulou, L. Bokobza, A. Kyritsis, V. Peoglos, P. Pissis, Comparative studies on effects of silica and titania nanoparticles on crystallization and complex segmental dynamics in poly(dimethylsiloxane). Polymer **51**, 5490–5499 (2010)

52. J.S. Meth, S.R. Lustig, Polymer interphase structure near nanoscale inclusions: comparison between random walk theory and experiment. Polymer **51**, 4259–4266 (2010)
53. I. Zaman, Q.-H. Le, H. Kuan, N. Kawashima, L. Luong, A.R. Gerson, J. Ma, Interface-tuned epoxy/clay nanocomposites. Polymer **52**, 497–504 (2011)
54. P.A. Klonos, C. Pandis, S. Kripotou, A. Kyritsis, P. Pissis, Interfacial effects in polymer nanocomposites studied by dielectric and thermal techniques. IEEE Trans. Dielectr. Electr. Insul. **19**, 1283–1290 (2012)
55. N. Jouault, M.K. Crawford, C. Chi, R. Smalley, B.A. Wood, J. Jestin, Y.B. Melnichenko, L. He, W.E. Guise, S.K. Kumar, Polymer chain behavior in polymer nanocomposites with attractive interactions. ACS Macro Lett. **5**, 523–527 (2016)
56. Y. Song, Q. Zheng, Concepts and conflicts in nanoparticles reinforcement to polymers beyond hydrodynamics. Prog. Mater. Sci. **84**, 1–58 (2016)
57. D. Wu, Y. Ge, R. Li, Y. Feng, P. Akcora, Thermally activated shear stiffening in polymer-grafted nanoparticle composites for high-temperature adhesives. ACS Appl. Polym. Mater. **4**, 2819–2827 (2022)
58. J. Huang, J. Zhou, M. Liu, Interphase in polymer nanocomposites. JACS Au **2**, 280–291 (2022)
59. B. Li, W. You, L. Peng, X. Huang, W. Yu, Revealing the shear effect on the interfacial layer in polymer nanocomposites through nanofiber reorientation. Macromolecules **56**, 3050–3063 (2023)
60. F.W. Starr, T.B. Schrøder, S.C. Glotzer, Molecular dynamics simulation of a polymer melt with a nanoscopic particle. Macromolecules **35**, 4481–4492 (2002)
61. M. Vacatello, Predicting the molecular arrangements in polymer-based nanocomposites. Macromol. Theory Simul. **12**, 86–91 (2003)
62. J.D. Smith, D. Bedrov, G.D. Smith, A molecular dynamics simulation study of nanoparticle interactions in a model polymer-nanoparticle composite. Compos. Sci. Technol. **63**, 1599–1605 (2003)
63. M.S. Ozmusul, C.R. Picu, S.S. Sternstein, S.K. Kumar, Lattice Monte Carlo simulations of chain conformations in polymer nanocomposites. Macromolecules **38**, 4495–4500 (2005)
64. S. Sen, J.D. Thomin, S.K. Kumar, P. Keblinski, Molecular underpinnings of the mechanical reinforcement in polymer nanocomposites. Macromolecules **40**, 4059–4067 (2007)
65. G. Allegra, G. Raos, M. Vacatello, Theories and simulations of polymer-based nanocomposites: from chain statistics to reinforcement. Prog. Polym. Sci. **33**, 683–731 (2008)
66. J.Y. Carrillo, S. Cheng, R. Kumar, M. Goswami, A.P. Sokolov, B.G. Sumpter, Untangling the effects of chain rigidity on the structure and dynamics of strongly adsorbed polymer melts. Macromolecules **48**, 4207–4219 (2015)
67. J.T. Kalathi, S.K. Kumar, M. Rubinstein, G.S. Grest, Rouse mode analysis of chain relaxation in polymer nanocomposites. Soft Matter. **11**, 4123–4132 (2015)
68. A.J. Trazkovich, M.F. Wendt, L.M. Hall, Effect of copolymer sequence on local viscoelastic properties near a nanoparticle. Macromolecules **52**, 513–527 (2019)
69. H. Reda, A. Chazirakis, A.F. Behbahani, N. Savva, V. Harmandaris, Revealing the role of chain conformations on the origin of the mechanical reinforcement in glassy polymer nanocomposites. Nano Lett. **24**, 148–155 (2024)
70. N. Jouault, D. Zhao, S.K. Kumar, Role of casting solvent on nanoparticle dispersion in polymer nanocomposites. Macromolecules **47**, 5246–5255 (2014)
71. G. Heinrich, M. Kluppel, T.A. Vilgis, Reinforcement of elastomers. Curr. Opin. Solid State Mater. Sci. **6**, 195–203 (2002)
72. J.L. Leblanc, Rubber-filler interactions and rheological properties in filled compounds. Prog. Polym. Sci. **27**, 627–687 (2002)
73. J.E. Mark, R. Abou-Hussein, T.Z. Sen, A. Kloczkowski, Some simulations on filler reinforcement in elastomers. Polymer **46**, 8894–8904 (2005)
74. S.Z. Wu, J.E. Mark, Some simulations and theoretical studies on poly(dimethylsiloxane). Polym. Rev. **47**, 463–485 (2007)

75. J. Liu, L.Q. Zhang, D. Cao, J. Shen, Y. Gao, Computational simulation of elastomer nanocomposites: current progress and future challenges. Rubber Chem. Technol. **85**, 450–481 (2012)
76. V.A. Buryachenko, A. Roy, K. Lafdi, K.L. Anderson, S. Chellapilla, Multi-scale mechanics of nanocomposites including interface: experimental and numerical investigation. Compos. Sci. Technol. **65**, 2435–2465 (2005)
77. D. Bedrov, G.D. Smith, J.S. Smith, Matrix-induced nanoparticle interactions in a polymer melt: a molecular dynamics simulation study. J. Chem. Phys. **119**, 10438–10447 (2003)
78. S.K. Kumar, R. Krishnamoorti, Nanocomposites: structure, phase behavior, and properties. Annu. Rev. Chem. Biomol. Eng. **1**, 37–58 (2010)
79. M. Aubouy, E. Raphaël, Scaling description of a colloidal particle clothed with polymers. Macromolecules **31**, 4357–4363 (1998)
80. J. Berriot, F. Lequeux, L. Monnerie, H. Montes, D. Long, P. Sotta, Filler-elastomer interaction in model filled rubbers, a ^{1}H NMR study. J. Non-Cryst. Solids **307–310**, 719–724 (2002)
81. L.M. Hall, A. Jayaraman, K.S. Schweizer, Molecular theories of polymer nanocomposites. Curr. Opin. Solid State Mater. Sci. **14**, 38–48 (2010)
82. H. Montes, T. Chaussée, A. Papon, F. Lequeux, L. Guy, Particles in model filled rubber: dispersion and mechanical properties. Eur. Phys. J. E **31**, 263–268 (2010)
83. G. Huber, T.A. Vilgis, Polymer adsorption on heterogeneous surfaces. Eur. Phys. J. B Condensed Matter Compl. Syst. **3**, 217–223 (1998)
84. C. Ohrt, T. Koschine, K. Rätzke, F. Faupel, L. Willner, G.J. Schneider, Free volume in PEP-silica nanocomposites with varying molecular weight. Polymer **55**, 143–149 (2014)
85. G.D. Smith, D. Bedrov, L. Li, O. Byutner, A molecular dynamics simulation study of the viscoelastic properties of polymer nanocomposites. J. Chem. Phys. **117**, 9478–9489 (2002)
86. T. Chen, H.-J. Qian, Y.-L. Zhu, Z.-Y. Lu, Structure and dynamics properties at interphase region in the composite of polystyrene and cross-linked polystyrene soft nanoparticle. Macromolecules **48**, 2751–2760 (2015)
87. M. Vacatello, Monte Carlo simulations of polymer melts filled with solid nanoparticles. Macromolecules **34**, 1946–1952 (2001)
88. D.C. Driscoll, H.S. Gulati, R.J. Spontak, C.K. Hall, Mixtures of polymer tails and loops grafted to an impenetrable interface. Polymer **39**, 6339–6346 (1998)
89. Y.N. Pandey, A. Brayton, C. Burkhart, G.J. Papakonstantopoulos, M. Doxastakis, Multiscale modeling of polyisoprene on graphite. J. Chem. Phys. **140**, 054908 (2014)
90. S. Rätzke, J. Kindersberger, The effect of interphase structures in nanodielectrics. IEEJ Trans. Fundam. Mater. **26**, 1044–1049 (2006)
91. X. Zhang, B.-W. Li, L. Dong, H. Liu, W. Chen, Y. Shen, C.-W. Nan, Superior energy storage performances of polymer nanocomposites via modification of filler/polymer interfaces. Adv. Mater. Interfaces **5**, 1800096 (2018)
92. S. Raetzke, J. Kindersberger, Role of interphase on the resistance to high-voltage arcing, on tracking and erosion of silicone/SiO_2 nanocomposites. IEEE Trans. Dielectr. Electr. Insul. **17**, 607–614 (2010)
93. J.C. Berg, Electrical properties of interfaces, in *An Introduction to Interfaces and Colloids. The Bridge to Nanoscience* (World Scientific, Hackensack, NJ, 2010), pp. 455–524
94. J. Lyklema, *Fundamentals of Interface and Colloid Science*, vol. II (Academic Press, San Diego, 1995)
95. T.J. Lewis, Interfaces are the dominant feature of dielectrics at the nanometeric level. IEEE Trans. Dilectr. Electr. Insul. **11**, 739–753 (2004)
96. M.A. Brown, Z. Abbas, A. Kleibert, R.G. Green, A. Goel, S. May, T.M. Squires, Determination of surface potential and electrical double-layer structure at the aqueous electrolyte-nanoparticle interface. Phys. Rev. X **6**, 011007 (2016)
97. A. Allagui, H. Benaoum, O. Olendski, On the Gouy-Chapman-Stern model of the electrical double-layer structure with a generalized Boltzmann factor. Physica A **582**, 126252 (2021)
98. K. Doblhoff-Dier, M.T.M. Koper, Modeling the Gouy-Chapman diffuse capacitance with attractive ion-surface interaction. J. Phys. Chem. C **125**, 16664–16673 (2021)

99. N. Jiang, M.K. Endoh, T. Koga, T. Masui, H. Kishimoto, M. Nagao, S.K. Satija, T. Taniguchi, Nanostructures and dynamics of macromolecules bound to attractive filler surfaces. ACS Macro Lett. **4**, 838–842 (2015)

100. N. Jouault, M.K. Crawford, C. Chi, R.J. Smalley, B. Wood, J. Jestin, Y.B. Melnichenko, L. He, W.E. Guise, S.K. Kumar, Polymer chain behavior in polymer nanocomposites with attractive interactions. ACS Macro Lett. **5**, 523–527 (2016)

101. H. Mortazavian, C.J. Fennell, F.D. Blum, Surface bonding is stronger for poly(methyl methacrylate) than for poly(vinyl acetate). Macromolecules **49**, 4211–4219 (2016)

102. H. Mortazavian, C.J. Fennell, F.D. Blum, Structure of the interfacial region in adsorbed poly(vinyl acetate) on silica. Macromolecules **49**, 298–307 (2016)

103. Y. Li, A.M. Waas, E.M. Arruda, A closed-form, hierarchical, multi-interphase model for composites - derivation, verification and application to nanocomposites. J. Mech. Phys. Solids **59**, 43–63 (2011)

104. H. Liu, L.C. Brinson, Reinforcing efficiency of nanoparticles: a simple comparison for polymer nanocomposites. Compos. Sci. Technol. **68**, 1502–1512 (2008)

105. A. Celzard, E. McRae, C. Deleuze, M. Dufort, G. Furdin, J.F. Marêché, Critical concentration in percolating systems containing a high-aspect-ratio-filler. Phys. Rev. B **53**, 6209–6214 (1996)

106. S.C. Baxter, B.J. Burrows, B.S. Fralick, Mechanical percolation in nanocomposites: microstructure and micromechanics. Probabilistic Eng. Mech. **44**, 35–42 (2016)

107. R.A. Vaia, J.F. Maguire, Polymer nanocomposites with prescribed morphology: going beyond nanoparticle-filled polymers. Chem. Mater. **19**, 2736–2751 (2007)

108. S.C. Baxter, C.T. Robinson, Pseudo-percolation: critical volume fractions and mechanical percolation in polymer nanocomposites. Compos. Sci. Technol. **71**, 1273–1279 (2011)

109. M. Mazaheri, J. Payandehpeyman, S. Jamasb, Modeling of effective electrical conductivity and percolation behavior in conductive-polymer nanocomposites reinforced with spherical carbon black. Appl. Compos. Mater. **29**, 695–710 (2022)

110. Q. Jiang, J. Yang, P. Hing, H. Ye, Recent advances, design guidelines, and prospects of flexible organic/inorganic thermoelectric composites. Mater. Adv. **1**, 1038–1054 (2020)

111. D. Stauffer, A. Aharony, *Introduction to Percolation Theory*, 2nd edn. (Taylor & Francis, London, 1992)

112. J.F. Gao, Z.M. Li, Q.J. Meng, Q. Yang, CNTs/UHMWPE composites with a two-dimensional conductive network. Mater. Lett. **62**, 3530–3532 (2008)

113. M. Meloni, M.J. Large, J.M. González Domínguez, S. Victor-Román, G. Fratta, E. Istif, O. Tomes, J.P. Salvage, C.P. Ewels, M. Pelaez-Fernandez, R. Arenal, A. Benito, W.K. Maser, A.A.K. King, P.M. Ajayan, S.P. Ogilvie, A.B. Dalton, Explosive percolation yields highly-conductive polymer nanocomposites. Nat. Commun. **13**, 6872 (2022)

114. W. Bauhofer, J.Z. Kovacs, A review and analysis of electrical percolation in carbon nanotube polymer composites. Compos. Sci. Technol. **69**, 1486–1498 (2009)

115. H.-J. Choi, M.S. Kim, D. Ahn, S.Y. Yeo, S. Lee, Electrical percolation threshold of carbon black in a polymer matrix and its application to antistatic fibre. Sci. Rep. **9**, 6338 (2019)

116. C.S. Boland, U. Khan, G. Ryan, S. Barwich, R. Charifou, A. Harvey, C. Backes, Z. Li, M.S. Ferreira, M.E. Möbius, R.J. Young, J.N. Coleman, Sensitive electromechanical sensors using viscoelastic graphene-polymer nanocomposites. Science **354**, 1257–1260 (2016)

117. S.Y. Kim, Y.J. Noh, J. Yu, Prediction and experimental validation of electrical percolation by applying a modified micromechanics model considering multiple heterogeneous inclusions. Compos. Sci. Technol. **106**, 156–162 (2015)

118. J. Du, L. Zhao, Y. Zeng, L. Zhang, F. Li, P. Liu, C. Liu, Comparison of electrical properties between multi-walled carbon nanotube and graphene nanosheet/high density polyethylene composites with a segregated network structure. Carbon **49**, 1094–1100 (2011)

119. X.J. He, J.H. Du, Z. Ying, H.M. Cheng, Positive temperature coefficient effect in multiwalled carbon nanotube/high-density polyethylene composites. Appl. Phys. Lett. **86**, 062112 (2005)

120. Q.H. Zhang, S. Rastogi, D.J. Chen, D. Lippits, P.J. Lemstra, Low percolation threshold in single-walled carbon nanotube/high-density polyethylene composites prepared by melt processing technique. Carbon **44**, 778–785 (2006)

121. P. Ciselli, R. Zhang, Z. Wang, C.T. Reynolds, M. Baxendale, T. Peijs, Oriented UHMW-PE/CNT composite tapes by a solution casting-drawing process using mixed-solvents. Eur. Polym. J. **45**, 2741–2748 (2009)
122. D.S. McLachlan, M. Blaszkiewicz, R.E. Newnham, Electrical resistivity of composites. J. Am. Ceram. Soc. **73**, 2187–2203 (1990)
123. M. Niklaus, H.R. Shea, Electrical conductivity and Young's modulus of flexible nanocomposites made by metal-ion implantation of polydimethylsiloxane: the relationship between nanostructure and macroscopic properties. Acta Mater. **59**, 830–840 (2011)
124. N. Ouali, J.-Y. Cavaillé, J. Perez, Elastic viscoelastic and plastic behavior of multiphase polymer blends. Plast. Rubber Compos. **16**, 55–60 (1991)
125. G.D. Seidel, D.C. Lagoudas, A micromechanics model for the electrical conductivity of nanotube-polymer nanocomposites. J. Compos. Mater. **43**, 917–941 (2009)
126. H. Liu, L.C. Brinson, Reinforcing efficiency of nanoparticles: a simple comparison for polymer nanocomposites. Compos. Sci. Technol. **68**, 147–162 (2007)
127. B.S. Fralick, E.P. Gatzke, S.C. Baxter, Three-dimensional evolution of mechanical percolation in nanocomposites with random microstructures. Probabilistic Eng. Mech. **30**, 1–8 (2012)
128. R. Bourn, B.S. Fralick, S.C. Baxter, Distributions of elastic moduli in mechanically percolating composites. Probabilistic Eng. Mech. **34**, 67–72 (2013)
129. M. Paley, J. Aboudi, Micromechanical analysis of composites by the generalized cells models. Mech. Mater. **14**, 127–139 (1992)
130. J. Aboudi, S.M. Arnold, B.A. Bednarcyk, *Micromechanics of Composite Materials: A Generalized Multiscale Analysis Approach* (Elsevier, Amsterdam, 2013)
131. T. Mori, K. Tanaka, Average stress in matrix and average elastic energy of materials with mis-fitting inclusions. Acta Mettall. **2**, 571–574 (1973)
132. Y. Benveniste, A new approach to the application of Mori-Tanaka theory in composite materials. Mech. Mater. **6**, 147–157 (1987)
133. M. Mazaheri, J. Payandehpeyman, M. Khamehchi, A developed theoretical model for effective electrical conductivity and percolation behavior of polymer-graphene nanocomposites with various exfoliated filleted nanoplatelets. Carbon **169**, 264–275 (2020)
134. A.J. Marsden, D.G. Papageorgiou, C. Vallés, A. Liscio, V. Palermo, M.A. Bissett, R.J. Young, I.A. Kinloch, Electrical percolation in graphene–polymer composites. 2D Mater. **5**, 032003 (2018)
135. C. Gao, S. Zhang, F. Wang, B. Wen, C. Han, Y. Ding, M. Yang, Graphene networks with low percolation threshold in ABS nanocomposites: selective localization and electrical and rheological properties. ACS Appl. Mater. Interfaces **6**, 12252–12260 (2014)
136. P. Steurer, R. Wissert, R. Thomann, R. Mülhaupt, Functionalized graphenes and thermoplastic nanocomposites based upon expanded graphite oxide. Macromol. Rapid Commun. **30**, 316–327 (2009)
137. D. Zheng, G. Tang, H.B. Zhang, Z.Z. Yu, F. Yavari, N. Koratkar, S.H. Lim, M.W. Lee, In situ thermal reduction of graphene oxide for high electrical conductivity and low percolation threshold in polyamide 6 nanocomposites. Compos. Sci. Technol. **72**, 284–289 (2012)
138. B.J. Rashmi, K. Prashantha, M.F. Lacrampe, P. Krawczak, Scalable production of multifunctional bio-based polyamide 11/graphene nanocomposites by melt extrusion processes via masterbatch approach. Adv. Polym. Technol. **37**, 1067–1075 (2018)
139. M. Beckert, F.J. Tölle, B. Bruchmann, R. Mülhaupt, Nitrogen-doped multilayer graphene as functional filler for carbon/polyamide 12 nanocomposites. Macromol. Mater. Eng. **300**, 785–792 (2015)
140. C. Xu, J. Gao, H. Xiu, X. Li, J. Zhang, F. Luo, Q. Zhang, F. Chen, Q. Fu, Can in situ thermal reduction be a green and efficient way in the fabrication of electrically conductive polymer/reduced graphene oxide nanocomposites? Compos. A Appl. Sci. Manuf. **53**, 24–33 (2013)
141. M. Yoonessi, J.R. Gaier, Highly conductive multifunctional graphene polycarbonate nanocomposites. ACS Nano **4**, 7211–7220 (2010)

142. T. Nezakati, A. Tan, A.M. Seifalian, Enhancing the electrical conductivity of a hybrid POSS-PCL/graphene nanocomposite polymer. J. Colloid Interface Sci. **435**, 145–155 (2014)

143. C. Tu, K. Nagata, S. Yan, Influence of melt-mixing processing sequence on electrical conductivity of polyethylene/polypropylene blends filled with graphene. Polym. Bull. **74**, 1237–1252 (2017)

144. A.A. Vasileiou, M. Kontopoulou, A. Docoslis, A noncovalent compatibilization approach to improve the filler dispersion and properties of polyethylene/graphene composites. ACS Appl. Mater. Interfaces **6**, 1916–1925 (2014)

145. M. Castelain, G. Martinez, C. Marco, G. Ellis, H.J. Salavagione, Effect of click-chemistry approaches for graphene modification on the electrical, thermal, and mechanical properties of polyethylene/graphene nanocomposites. Macromolecules **46**, 8980–8987 (2013)

146. H. Hu, G. Zhang, L. Xiao, H. Wang, Q. Zhang, Z. Zhao, Preparation and electrical conductivity of graphene/ultrahigh molecular weight polyethylene composites with a segregated structure. Carbon **50**, 4596–4599 (2012)

147. M. Sabzi, L. Jiang, F. Liu, I. Ghasemi, M. Atai, Graphene nanoplatelets as poly(lactic acid) modifier: linear rheological behavior and electrical conductivity. J. Mater. Chem. A **1**, 8253–8261 (2013)

148. J.H. Yang, S.H. Lin, Y.D. Lee, Preparation and characterization of poly(l-lactide)-graphene composites using the in situ ring-opening polymerization of PLLA with graphene as the initiator. J. Mater. Chem. **22**, 10805–10815 (2012)

149. X. Zeng, J. Yang, W. Yuan, Preparation of a poly(methyl methacrylate)-reduced graphene oxide composite with enhanced properties by a solution blending method. Eur. Polym. J. **48**, 1674–1682 (2012)

150. B. Mutlay, L.B. Tudoran, M. Ibrahim, L.B. Tudoran, Percolation behavior of electrically conductive graphene nanoplatelets/polymer nanocomposites: theory and experiment. Fuller. Nanotub. Carbon Nanostructures **22**, 413–433 (2014)

151. H.B. Zhang, W.G. Zheng, Q. Yan, Z.G. Jiang, Z.Z. Yu, The effect of surface chemistry of graphene on rheological and electrical properties of polymethylmethacrylate composites. Carbon **50**, 5117–5125 (2012)

152. Y. Lan, H. Liu, X. Cao, S. Zhao, K. Dai, X. Yan, G. Zheng, C. Liu, C. Shen, Z. Guo, Electrically conductive thermoplastic polyurethane/polypropylene nanocomposites with selectively distributed graphene. Polymer **97**, 11–19 (2016)

153. D. Hofmann, K.A. Wartig, R. Thomann, B. Dittrich, B. Schartel, R. Mulhaupt, Functionalized graphene and carbon materials as additives for melt-extruded flame retardant polypropylene. Macromol. Mater. Eng. **298**, 1322–1334 (2013)

154. W. Fan, C. Zhang, W.W. Tjiu, T.X. Liu, Fabrication of electrically conductive graphene/polystyrene composites via a combination of latex and layer-by-layer assembly approaches. J. Mater. Res. **28**, 611–619 (2013)

155. N. Liu, F. Luo, H. Wu, Y. Liu, C. Zhang, J. Chen, One-step ionic-liquid-assisted electrochemical synthesis of ionic-liquid-functionalized graphene sheets directly from graphite. Adv. Funct. Mater. **18**, 1518–1525 (2008)

156. X.Y. Qi, D. Yan, Z. Jiang, Y.K. Cao, Z.Z. Yu, F. Yavari, N. Koratkar, Enhanced electrical conductivity in polystyrene nanocomposites at ultra-low graphene content. ACS Appl. Mater. Interfaces **3**, 3130–3133 (2011)

157. P. Zhao, Y. Luo, J. Yang, D. He, L. Kong, P. Zheng, Q. Yang, Electrically conductive graphene-filled polymer composites with well organized three-dimensional microstructure. Mater. Lett. **121**, 74–77 (2014)

158. Y.C. Chiu, C.L. Huang, C. Wang, Rheological and conductivity percolations of syndiotactic polystyrene composites filled with graphene nanosheets and carbon nanotubes: A comparative study. Compos. Sci. Technol. **134**, 153–160 (2016)

159. Z. Tu, J. Wang, C. Yu, H. Xiao, T. Jiang, Y. Yang, D. Shi, Y.W. Mai, R.K.Y. Li, A facile approach for preparation of polystyrene/graphene nanocomposites with ultra-low percolation threshold through an electrostatic assembly process. Compos. Sci. Technol. **134**, 49–56 (2016)

160. A.K. Appel, R. Thomann, R. Mulhaupt, Polyurethane nanocomposites prepared from solvent-free stable dispersions of functionalized graphene nanosheets in polyols. Polymer **53**, 4931–4939 (2012)
161. N. Yousefi, M.M. Gudarzi, Q. Zheng, S.H. Aboutalebi, F. Sharif, J.K. Kim, Self-alignment and high electrical conductivity of ultralarge graphene oxide-polyurethane nanocomposites. J. Mater. Chem. **22**, 12709–12717 (2012)
162. S. Vadukumpully, J. Paul, N. Mahanta, S. Valiyaveettil, Flexible conductive graphene/poly(vinyl chloride) composite thin films with high mechanical strength and thermal stability. Carbon **49**, 198–205 (2011)
163. J.W. Zha, B. Zhang, R.K.Y. Li, Z.M. Dang, High-performance strain sensors based on functionalized graphene nanoplates for damage monitoring. Compos. Sci. Technol. **123**, 32–38 (2016)
164. Y. Li, J. Tang, L. Huang, Y. Wang, J. Liu, X. Ge, S.C. Tjong, R.K.Y. Li, L.A. Belfiore, Facile preparation, characterization and performance of noncovalently functionalized graphene/epoxy nanocomposites with poly(sodium 4-styrenesulfonate). Compos. A Appl. Sci. Manuf. **68**, 1–9 (2015)
165. R. Moriche, M. Sánchez, A. Jiménez-Suárez, S.G. Prolongo, A. Ureña, Electrically conductive functionalized-GNP/epoxy based composites: from nanocomposite to multiscale glass fibre composite material. Compos. B Eng. **98**, 49–55 (2016)
166. M. Monti, M. Rallini, D. Puglia, L. Peponi, L. Torre, J.M. Kenny, Morphology and electrical properties of graphene-epoxy nanocomposites obtained by different solvent assisted processing methods. Compos. A Appl. Sci. Manuf. **46**, 166–172 (2013)
167. Q. Meng, H. Wu, Z. Zhao, S. Araby, S. Lu, J. Ma, Free-standing, flexible, electrically conductive epoxy/graphene composite films. Compos. A Appl. Sci. Manuf. **92**, 42–50 (2017)
168. K. Tschoppe, F. Beckert, M. Beckert, R. Mülhaupt, Thermally reduced graphite oxide and mechanochemically functionalized graphene as functional fillers for epoxy nanocomposites. Macromol. Mater. Eng. **300**, 140–152 (2015)
169. Y. Li, H. Zhang, H. Porwal, Z. Huang, E. Bilotti, T. Peijs, Mechanical, electrical and thermal properties of in-situ exfoliated graphene/epoxy nanocomposites. Compos. A Appl. Sci. Manuf. **95**, 229–236 (2017)
170. S. Zhao, H. Chang, S. Chen, J. Cui, Y. Yan, High-performance and multifunctional epoxy composites filled with epoxide-functionalized graphene. Eur. Polym. J. **84**, 300–312 (2016)
171. S. Wu, R.B. Ladani, J. Zhang, E. Bafekrpour, K. Ghorbani, A.P. Mouritz, A.J. Kinloch, C.H. Wang, Aligning multilayer graphene flakes with an external electric field to improve multifunctional properties of epoxy nanocomposites. Carbon **94**, 607–618 (2015)
172. G. Tang, Z.G. Jiang, X. Li, H.B. Zhang, S. Hong, Z.Z. Yu, Electrically conductive rubbery epoxy/diamine-functionalized graphene nanocomposites with improved mechanical properties. Compos. B Eng. **67**, 564–570 (2014)
173. O.K. Park, J.Y. Hwang, M. Goh, J.H. Lee, B.C. Ku, N.H. You, Mechanically strong and multifunctional polyimide nanocomposites using amimophenyl functionalized graphene nanosheets. Macromolecules **46**, 3505–3511 (2013)
174. O.K. Park, S.G. Kim, N.H. You, B.C. Ku, D. Hui, J.H. Lee, Synthesis and properties of iodo functionalized graphene oxide/polyimide nanocomposites. Compos. B Eng. **56**, 365–371 (2014)
175. Y. Zhan, M. Lavorgna, G. Buonocore, H. Xia, Enhancing electrical conductivity of rubber composites by constructing interconnected network of self-assembled graphene with latex mixing. J. Mater. Chem. **22**, 10464–10468 (2012)
176. J.R. Potts, O. Shankar, L. Du, R.S. Ruoff, Processing-morphology-property relationships and composite theory analysis of reduced graphene oxide/natural rubber nanocomposites. Macromolecules **45**, 6045–6055 (2012)
177. B. Dong, S. Wu, L. Zhang, Y. Wu, High performance natural rubber composites with well-organized interconnected graphene networks for strain-sensing application. Ind. Eng. Chem. Res. **55**, 4919–4929 (2016)

Chapter 4
Mechanical and Acoustic Characteristics of Polymer Nanocomposites

The polymer nanocomposites (PNCs) filled with either graphene particles or its derivatives attract considerable interest due to possibility to tailor physical and chemical properties of PNCs encountered with only a small quantity of nanofiller incorporated to the host polymer matrix. Numerous studies have shown that polymer nanocomposites filled with single-layer graphene nanosheets (SLG), multi-layered or platelet graphene (MLG), as well as their oxides and chemically modified derivatives exhibit substantial property enhancements at much lower loadings than with other conventional nanofillers. On the other hand, it is recognized in the literature, that the overall physical and chemical behavior of PNCs is significantly influenced not only by intrinsic properties, geometry, and spatial distribution of embedded nanoparticles but also by the presence of so-called interphase layers (IPLs) arising in the vicinity of nanoparticles. In particular, IPLs play an important role in governing the stress transfer over polymer–nanofiller interface and, thus, in controlling the failure mechanisms and fracture toughness of a PNC. The size of the interphase region can be discovered from ultrasonic measurements, thereby acoustic wave propagation methods are of great importance in polymer–graphene nanocomposites experimental investigations.

4.1 Mechanical Properties of Polymers

The mechanical properties of polymers are most often obtained using a uniaxial tensile test at a constant rate of strain or head motion similar to those used for metals and other materials [1]. Schematic stress–strain ($T-S$) diagrams characteristic of those found for the indicated types of solid polymers are shown in Fig. 4.1. Curve 1 represents a linear elastic and brittle material like epoxy, polystyrene, etc. Curve 2 is obtained in a semi-ductile material like poly(methyl methacrylate). Curve 3 corresponds to a ductile material like polyethyleneimine or polycarbonate. Finally,

A. Nadtochiy et al., *Graphene-Based Polymer Nanocomposites*,
SpringerBriefs in Applied Sciences and Technology,
https://doi.org/10.1007/978-981-97-2792-6_4

Fig. 4.1 Typical
stress–strain diagrams of
various polymer types

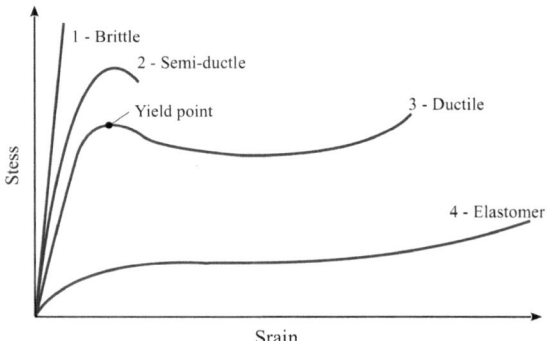

curve 4 describes the stress–strain behavior in a typical elastomer such as flexible urethane. The yield stress of a ductile material is approximated by the proportional limit stress or the first peak in the stress–strain diagram in Fig. 4.1 (termed the intrinsic yield point).

The response of polymer-based materials to stress is time dependent. One of the methods used to characterize the viscoelastic time-dependent behavior of a polymer is the relaxation test. In this technique, a constant strain is applied quasi-statically to a uniaxial tensile (compression or torsion) bar at zero time. The bar is suddenly stretched to a new position and rigidly fixed such that the strain remains constant for the duration of the test. If the stress is a function of time, $T(t)$, and the strain S is constant ($S = S_0$), the modulus $E(t)$ will also vary with time.

The resulting modulus $E(t)$ defined as the relaxation modulus of the polymer is given by

$$E(t) = \frac{T_0}{S(t)} \qquad (4.1)$$

or

$$T(t) = S_0 E(t). \qquad (4.2)$$

The latter equation is the uniaxial stress–strain relation for a polymer. It is similar to Hooke's law for a material that is time–independent but is valid only in the case of a constant input of stress. The relaxation test provides the equation for the material property termed as the relaxation modulus.

Another characterization test for viscoelastic materials is the creep test in which a uniaxial tensile (compression or torsion) bar is loaded with a constant stress at zero time. Again, the load is applied quasi-statically to avoid inertia effects, and the material is assumed to have no prior history. In this case, the strain under the constant load increases with time, and the test defines a quantity $D(t)$ called the creep compliance,

$$D(t) = \frac{S_0}{T(t)}. \qquad (4.3)$$

In this case,

$$S(t) = T_0 D(t). \tag{4.4}$$

In a creep test, the strain will tend to a constant value after a long time for a thermoset while the strain will increase without bound for a thermoplastic.

4.1.1 Phenomenological Mechanical Models

Consider elementary mechanical models that can describe some aspects of viscoelastic polymeric behavior [1]. These simple models cannot represent the behavior of real polymers over their complete history of use. Instead, they are very helpful in gaining physical insight into the phenomena of creep, relaxation, and other test procedures. They moreover provide a better understanding of the stress–strain relationship in a viscoelastic material.

The simplest mechanical models for viscoelastic behavior consist of two elements, (i) a spring for elastic behavior and (ii) a dashpot for viscous behavior; see Fig. 4.2a, b. In an ideal lossless medium, Hooke's law corresponds to the stress–strain relation for a linear spring

$$T = ES \tag{4.5}$$

where E is Young's modulus. For the dashpot, the relationship is

$$T = \eta \frac{\partial S}{\partial t} \tag{4.6}$$

where η is the viscosity. Spring and dashpot elements can be combined in a variety of arrangements to produce a simulated viscoelastic response. The elements can be combined in series. The combination is known as a Maxwell element and is illustrated in Fig. 4.2c. The combination of the elastic and viscous elements in parallel is known

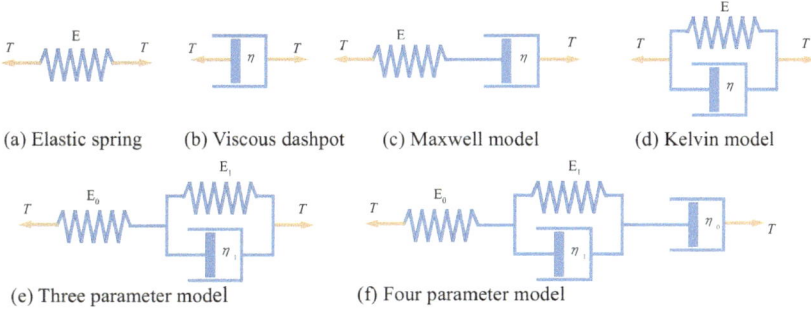

(a) Elastic spring   (b) Viscous dashpot   (c) Maxwell model   (d) Kelvin model

(e) Three parameter model   (f) Four parameter model

Fig. 4.2 Schematics showing elastic spring and viscous dashpot elements in series and parallel configurations for various models of linear viscoelasticity

as a Kelvin element (sometimes as a Voigt or Kelvin–Voigt element) and also is illustrated in Fig. 4.2d. These models are very useful in understanding the physical relation between stress and strain that occurs in polymers and other viscoelastic materials. For example, if suddenly a constant stress is applied as in a creep test, each model with a free spring will have a sudden increase in strain. The Kelvin will not have a sudden increase in strain as the damper will not allow a sudden jump in strain. Under the condition of constant stress, each model with a free damper (Maxwell and four-parameter fluid) will have an ever-increasing creep strain and will be similar to the response for a thermoplastic polymer. In a creep test, both the Kelvin and the three-parameter solid will creep to a limiting strain because the damper in each is constrained by the spring and as a result, the response will be similar to that of a thermoset polymer. In relaxation, the stress will decay to zero for those models with a free damper (Maxwell and four-parameter fluid) and the stress will decay to a limiting value for those without a free damper (Kelvin and three-parameter solid) for thermoplastic and thermosetting materials respectively. Note that a simple stress relaxation test is not possible for a Kelvin model as the damper will prohibit a sudden increase in strain. The above methods can be used to determine the differential equations, solutions, and parameters for several mechanical models using a variety of combinations of springs and damper elements. Table 4.1 is a tabulation of the differential equation, parameter inequalities, creep compliances, and relaxation moduli for frequently discussed basic models.

4.1.2   Acoustic Waves in Viscoelastic Materials

There are various schemes for deriving the equations of motion for structural elements. The approach taken here is to begin with the simplest available model, and then as the need arises to append modifications to it. The elementary theory considers the rod to be long and slender and assumes it supports only one-dimensional axial stress. It further assumes that the lateral contraction (or the Poisson's ratio effect) can be neglected. Let $q(x, t)$ be the externally applied body force per unit volume and $u(x, t)$ be the displacement in the x direction. Then, with reference to Fig. 4.3, the balance of forces gives

Table 4.1 Common linear viscoelastic models and their corresponding differential equations and modulus [2]

Model name	Differential equation	Complex modulus
Elastic	$T = ES$	E
Viscous	$T = \eta \dot{S}$	$i\omega\eta$
Maxwell	$\frac{T}{\eta} + \frac{\dot{T}}{E} = \dot{S}$	$\frac{\omega^2\eta^2 E}{E^2+\omega^2\eta^2} + j\frac{\omega\eta E}{E^2+\omega^2\eta^2}$
Kelvin–Voigt	$T = ES + \eta\dot{S}$	$E + j\omega\eta$

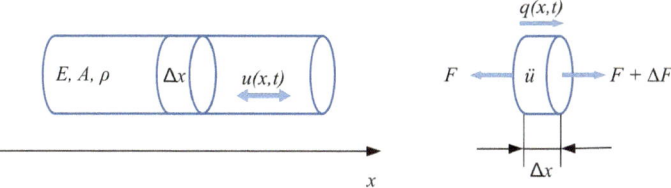

Fig. 4.3 Element of rod with loads

$$- F + (F + \Delta F) + q A \Delta x = \rho A \Delta x \ddot{u}$$

where ρA is the mass density per unit length of the rod. If the Δx quantities are very small, then the equation of motion becomes

$$\frac{\partial F}{\partial x} = \rho A \frac{\partial^2 u}{\partial t^2} - q A \tag{4.7}$$

The independent variables are x and t. It is now desirable to write the equation only in terms of the displacement. To do this, first consider the strain–displacement relation

$$e = \frac{\partial u}{\partial x} \tag{4.8}$$

and assume the material behavior to be linear elastic in the one-dimensional form

$$\frac{F}{A} = \sigma = Ee \tag{4.9}$$

where E = Young's Modulus. Then combining with the above gives

$$\frac{\partial}{\partial x} \left\{ E A \frac{\partial u}{\partial x} \right\} = \rho A \frac{\partial^2 u}{\partial t^2} - q A \tag{4.10}$$

If the stress (or strain) is taken as the dependent variable, it also would have an equation of similar form. That is, all dependent variables have an equation of the form

$$v_0^2 \frac{\partial^2 u}{\partial x^2} - \frac{\partial^2 u}{\partial t^2} = 0, \ v_0 = \sqrt{\frac{E}{\rho}} \tag{4.11}$$

Viscoelastic properties are often determined by complex Young's modulus $E^* = E - jE''$ [2]. The equation of motion (4.10) is given by

$$E^* \frac{\partial^2 u}{\partial x^2} = \rho \frac{\partial^2 u}{\partial t^2}, \tag{4.12}$$

where ρ is density. The solution of (4.12) is

$$u = u_0 e^{(j\omega t - \gamma x)}, \tag{4.13}$$

where γ is the propagation constant, ω is the angular frequency and j is the imaginary unit. γ is a complex number $\gamma = \alpha(\omega) + j\beta(\omega)$, where α is the attenuation constant and $\beta = \omega/v(\omega)$ is the phase shift constant and $v(\omega)$ is the phase velocity.

Materials with frequency-dependent elastic moduli are called dispersive. Measurements and theory show that sound absorption mechanisms lead to dispersion. The real and imaginary parts of an elastic modulus are related by the Kramers–Kronig relations which can be applied to acoustical systems concerning the linearity and causality conditions. Under the causality condition, the effect cannot precede the cause [3].

Relationships between attenuation and dispersion expressed in the form suitable for ultrasonic studies can be written as [4, 5]

$$K_1(\omega) - K_1(\infty) = \frac{2}{\pi} \int_0^\infty \frac{\omega' K_2(\omega')}{\omega'^2 - \omega^2} d\omega', \tag{4.14}$$

$$K_2(\omega) = -\frac{2}{\pi} \int_0^\infty \frac{K_1(\omega') - K_1(\infty)}{\omega'^2 - \omega^2} d\omega', \tag{4.15}$$

where $K_1(\omega)$ and $K_2(\omega)$ are the real and imaginary parts, respectively, of the dynamic compressibility (inverse of the bulk modulus). In the limit of $\alpha(\omega) v(\omega) / \omega \ll 1$, when the real part of the wave vector is much larger than its imaginary part, the real and imaginary parts of the compressibility can be directly related to the attenuation coefficient and phase velocity as

$$v(\omega) = \frac{1}{\sqrt{\rho K_1(\omega)}}, \tag{4.16}$$

$$\alpha(\omega) = \frac{1}{2}\rho v(\omega) K_2(\omega). \tag{4.17}$$

The significant feature of these relationships is that they do not depend upon details of the specific mechanism responsible for the sound attenuation and dispersion.

The relationship between shear wave attenuation and phase velocity data in an epoxy polymer was examined in the range over 5 decades of frequency [6]. It was shown that the attenuation characteristic is close to a straight line, so that $\alpha(f) = af^b$, where a and b are constants, $\alpha(f)$ is the attenuation measured in Nepers per meter (Np/m), and f is the sound frequency in Hertz (Hz). The continuous line on the attenuation graph corresponds to values of $\alpha = 340 \times 10^{-6}\,\mathrm{Np\,m^{-1}\,Hz^{-1}}$ and $b = 1.007$. The relationship between shear wave attenuation and phase velocity data in an epoxy polymer was examined in the range over 5 decades of frequency [6]. It was shown that the attenuation characteristic is close to a straight line, so that

$\alpha(f) = af^b$, where a and b are constants, $\alpha(f)$ is the attenuation measured in Nepers per meter (Np/m), and f is the sound frequency in Hertz (Hz). The continuous line on the attenuation graph corresponds to values of $\alpha = 340 \times 10^{-6}$ Np m^{-1} Hz^{-1} and $b = 1.007$. This value of b is very close to unity resulting in the continuous line on the phase velocity curve. O'Donnell et al. [5] have shown that given a representation of ultrasonic wave attenuation in the frequency domain $\alpha(\omega)$, then the phase velocity at any frequency can be calculated using the near local forms of the Kramers–Kronig relationships

$$\alpha(\omega) = \frac{\pi \omega^2}{2v^2(\omega)} \frac{dv(\omega)}{d\omega}, \tag{4.18}$$

$$\frac{1}{v(\omega_0)} - \frac{1}{v(\omega_1)} = \frac{2}{\pi} \int_{\omega_0}^{\omega_1} \frac{\alpha(\omega')}{\omega'^2} d\omega'. \tag{4.19}$$

Here, $v(\omega_0)$ is some value of the phase velocity at a low frequency ω_0 and $v(\omega_1)$ is the phase velocity that we wish to calculate at the higher frequency ω_1, given knowledge of the attenuation function. If one assumes proportionality between attenuation and frequency (i.e., $b = 1$) (4.19) becomes

$$\frac{1}{v(\omega_0)} - \frac{1}{v(\omega_1)} = \frac{a}{\pi^2} \left[ln \frac{\omega_1}{\omega_0} \right]. \tag{4.20}$$

This expression can be used to calculate the sound phase velocity as a function of frequency. The attenuation appears to be a linear function of frequency across the frequency band, and the measured phase velocity dispersion shows an agreement with the calculations to an accuracy within 10%.

4.2 Continuum Mechanics Models of Nanocomposites

Continuum mechanics can be employed to describe the mechanical behavior of molecular systems. As they have a discrete (not continuous) structure, the most favored model is the equivalent-continuum model [7]. In this model, the generalized constitutive equation of the equivalent continuum is

$$T_{ij} = C_{ijkl} S_{kl}, \tag{4.21}$$

where T_{ij} are the components of the stress tensor with $i, j = 1, 2, 3$, C_{ijkl} are the components of the linear-elastic stiffness tensor, and S_{kl} are the components of the strain tensor. Here, the summation over repeated subscript indices is assumed. Furthermore, it is assumed that the composite is isotropic because of spherical reinforcement.

4.2.1 Mori–Tanaka Model

Within the equivalent-continuum model formalism, two micromechanics techniques were proposed to describe the bulk elastic properties of composites. In the *Mori–Tanaka approach* [8], only the two phases, matrix and effective particles, are considered which are perfectly bonded to each other; see Fig. 4.4a. This technique was used to predict the elastic properties of two-phase composites as a function of the effective particle volume fraction and geometry.

For the Mori–Tanaka method, the overall elastic stiffness tensor of the composite containing the isotropic constituents is

$$\mathbf{C} = \frac{c^m \mathbf{C}^m + c^p \mathbf{C}^p \mathbf{S}^p}{c^m \mathbf{I} + c^p \mathbf{S}^p}, \tag{4.22}$$

where the terms with boldface refer to the tensor quantities, c^p and c^m are the effective particle and matrix volume fractions, $\mathbf{C}^p$ and $\mathbf{C}^m$ are the stiffness tensors of the effective particle and matrix, respectively, $\mathbf{I}$ is the identity tensor. The dilute strain-concentration tensor $\mathbf{S}^p$ of the effective particles, which relates local and global strains in the individual phase (filler particle, matrix, or interphase), is

Fig. 4.4 Schematics of the Mori–Tanaka (**a**) and effective interface micromechanics (**b**) models

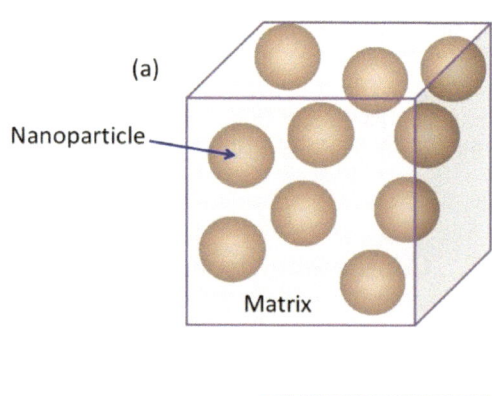

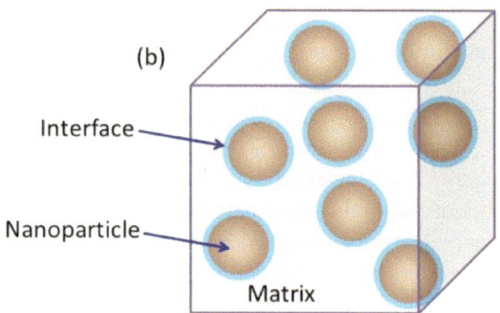

$$\mathbf{S}^p = \frac{1}{\mathbf{I} + \frac{\Xi^p}{\mathbf{C}^m}(\mathbf{C}^p - \mathbf{C}^m)}, \tag{4.23}$$

where Ξ^p is the Eshelby tensor.

Based on the stress symmetry condition, the four full subscripts in the stiffness and Eshelby tensor components may be reduced to two abbreviated subscripts by using abbreviated subscript notation $ijkl \leftrightarrow \alpha\beta$, where α is related to (ij), and β to (kl) as

$$(11) \leftrightarrow 1, (22) \leftrightarrow 2, (33) \leftrightarrow 3,$$
$$(23) = (32) \leftrightarrow 4, (13) = (31) \leftrightarrow 5, (12) = (21) \leftrightarrow 6. \tag{4.24}$$

Then, for spherical effective filler particles and an isotropic matrix, the components of the Eshelby tensor in the notations of (4.24) are [9, 10]

$$\Xi_{11} = \Xi_{22} = \Xi_{33} = \frac{7 - 5\nu}{15(1 - \nu)},$$
$$\Xi_{12} = \Xi_{13} = \Xi_{23} = \Xi_{32} = \Xi_{31} = \Xi_{21} = \frac{5\nu - 1}{15(1 - \nu)}$$
$$\Xi_{44} = \Xi_{55} = \Xi_{66} = \frac{4 - 5\nu}{15(1 - \nu)}, \tag{4.25}$$

where ν is the Poisson ratio of the matrix material. It is clear from (4.23)–(4.25) that the composite stiffness tensor is isotropic. For composites with non-spherical reinforcement, (4.25) has a different form, and the $C_{\alpha\beta}$ tensor is generally not tractable in the isotropic form.

The Mori–Tanaka approach has been successively used for tailoring the overall properties of composites when the reinforcements are on a length scale larger than micrometers. However, for nanometer-sized reinforcement, the restriction of the two phases is deemed inappropriate for reasons that the molecular structure of the polymer matrix is significantly perturbed at the reinforcement/polymer interface. In this case, the perturbed region is on a length scale of the nanometer-sized reinforcement [9].

4.2.2 Effective Interface Model

Another approach to improving the utility of the equivalent-continuum model is to include the interface region in the *effective interface model* [11] as shown in Fig. 4.4b. The effective interface has a finite size and can be referred to as the above interphase zone. For this model, the bulk elastic stiffness tensor may be represented as

$$\mathbf{C} = \mathbf{C}^m + \frac{(c^p + c^i)(\mathbf{C}^i - \mathbf{C}^m)\mathbf{S}^{pi} + c^p(\mathbf{C}^p - \mathbf{C}^i)\mathbf{S}^p}{c^m\mathbf{I} + (c^p + c^i)\mathbf{S}^{pi}}, \tag{4.26}$$

where c^i and $\mathbf{C}^i$ are the volume fraction and stiffness tensor for the interface, respectively, and $\mathbf{S}^p$ and $\mathbf{S}^{pi}$ are the dilute strain-concentration tensors given by

$$\mathbf{S}^p = \mathbf{I} - \frac{\Xi^p}{\Xi^p + \frac{\mathbf{C}^m}{\mathbf{C}^p - \mathbf{C}^m}},$$

$$\mathbf{S}^{pi} = \mathbf{I} - \Xi^p \left[\frac{c^p}{c^p + c^i} \frac{1}{\Xi^p + \frac{\mathbf{C}^m}{\mathbf{C}^p - \mathbf{C}^m}} + \frac{c^i}{c^p + c^i} \frac{1}{\Xi^p + \frac{\mathbf{C}^m}{\mathbf{C}^i - \mathbf{C}^m}} \right], \qquad (4.27)$$

where the Eshelby tensor Ξ^p is given by (4.25). Similar to the Mori–Tanaka model, the effective interface model predicts the composite stiffness tensor in (4.26) is isotropic for spherical particle reinforcement. In practice, the parameters of the model have to be adjusted to the polymer and filler materials.

4.3 Interfacial Characteristics of Polymer–Carbon Nanocomposites

The in-plane elastic modulus of pristine, defect-free graphene is very high, up to 1.1 TPa [12]. However, it has a large production cost, so that MLG thin films are primarily employed in fabricating nanocomposites [13]. Typically, the defect concentration in MLGs is rather high, which reduces the mechanical strength of fillers, but frequently yields functionalization and bonding to the polymeric matrix. Thus, it has been demonstrated that thermal stability and thermal conductivity can be simultaneously improved in epoxy nanocomposites filled with bare and oxidized MLGs [14]. The involvement of the interfacial interaction mechanisms in the thermal improvements and thermal stability was also discussed [15, 16].

In the electrical industry, there are also increasing requirements for high electrical performance and mechanical strength in polymer composites. Insulating polymeric materials can be made conductive by adding conventional conducting fillers and forming percolating networks, which, however, results in decreasing mechanical strength [17].

As discussed above, the interface conditions are crucial for the filler-based reinforcement. Incidentally, the pristine graphene does not form homogeneous composites as it is not compatible with organic polymers. Even if the interface layer between the filler particle and polymeric matrix is very thin, it is crucially important in controlling the nanocomposite properties, as the total surface area of nanofillers is high. Among different interface effects in nanocomposites [18, 19], the matrix/filler interface condition has a profound effect on the Young modulus, yield strength, creep, stress relaxation, and sensitivity of nanocomposites [17, 20–22]

Comparisons of micro-mechanical theoretical computations with experiments showed that reinforcement could mainly arise from the native filler properties [23]. In contrast, it was suggested that interface regions could be significant in explaining the mechanical reinforcement because a large volume fraction of the matrix is modified in nanocomposites [24]. Nanofillers can polarize the outer polymeric medium, which thus alters the molecular packing and dynamics near the filler and, indirectly, the ones far away from the filler. These can lead to highly non-local coupling of the stress and strain tensors [24].

Microscopically, the observed increase in the elastic modulus of polymeric media filled with a relatively small amount of fillers can be ascribed to the molecular stiffening [25]. This effect is proportional to the specific filler/matrix interface area (S_f) and the strength of the filler/matrix interactions. The model of trapped entanglements [25] can quantitatively describe the observed increase of elastic modulus due to molecular stiffening in the neat matrix with nanofillers.

The challenge for modeling polymer nanocomposite structures is to predict accurately their behavior in external fields and to capture the phenomena on length scales that span typically 5–6 orders of magnitude and time scales that can span a dozen orders of magnitude [26]. From this point of view, new strategies for multiscale modeling (MSM) and simulation have been proposed in the last decades to predict accurately the physical/chemical properties of nanocomposite materials. The strategies enable to bridge/link the models and simulation techniques across a broad range of lengths and time scales to address the macroscopic or mesoscopic behaviors of polymer nanocomposites, starting from a detailed description at the molecular/atomistic level of local motions of individual moieties and short chains [26, 27]. There are two basic multiscale approaches: (a) sequential or hierarchical methods and (b) hybrid or concurrent methods [27]. Hybrid methods seek to incorporate aspects of various size scale phenomena in a single simulation [28–30] In hierarchical modeling, first simulations at the higher resolution are performed and properties extracted are used as input in the next level method [31]. In recent hierarchical modeling has been extensively used for predicting mechanical properties of clay-filled polymer nanocomposites [32, 33] and nanocomposites filled with different carbon structures including CNT [34, 35], fuzzy carbon fibers [36, 37], graphene [38].

However, several micro- and mesoscale effects associated with long-range interactions, which are reminiscent of the enhanced interphase layer thickness in polymer nanocomposite structures have not yet been incorporated into the modeling schemes. Meanwhile, the long-range interactions play a major role in liquid reactive polymer-filler mixtures during the curing process. They are associated, in particular, with an action of electrostatic forces between the active charged surface sites of dispersed nanoparticles and dipole segments of macromolecules, and also with viscosity forces appearing between particle's surface and macromolecular chains [39], while both are captured in diffusion or convective mass-transfer microstreams. These microstreams must facilitate the Brownian motion of nanoparticles. The related effects include spatial density nonuniformity (referred to as the nodular structure) of polymeric matrices themselves [40, 41], spatial nonuniformities arising from the Brownian motion of nanoparticles in liquid reactive mixtures or melts [41, 42] and formation of double electric layers around the nanoparticles [43] due to the differences in the electrical properties of the embedded nanoparticles and a host polymer matrix.

Moreover, it is in the case of graphene-filled polymer nanocomposites, the heterogeneity of the distribution of bound charge density in the vicinity of nanoparticles can be further enhanced due to the morphological peculiarities of the latter. The matter is that the individual layers of graphene undergo out-of-plane wrapping, rippling, folding, scrolling, and crumpling [44] . Also, graphene nanoplatelets possess structural defects [45] . Further, theoretical calculations [46] shows that strain gradients imposed on curved graphene nanoplatelets or on those containing structural defects as noncentrosymmetric pores can accompanied by asymmetric redistribution of the electron density, which, in turn, induces strong polarization of the nanoplatelets. Thus, it may be assumed that in graphene-filled polymer nanocomposites, the above-mentioned phenomena can contribute to increasing the interphase layer thickness.

Also, earlier experimental studies [47, 48] show that for epoxy resins cured on metal (i.e., highly conductive) surfaces, interphase thicknesses can vary from several nanometers up to the micrometer range. It is pointed out in [49] that a stronger extension of the interphase zone on polymer-solid interfaces can be promoted by lower linker concentration in the reactive mixture.

Mechanical properties and propagation of elastic waves in composite materials containing accidentally dispersed particles have been extensively addressed both theoretically [50] and experimentally [51]. One area of theoretical studies deals with dispersion relations of elastic waves in two-phase composites when multiple scattering on immersed particles of spherical [50] or cylindrical [50] shapes are taken into account. Mass density, compression, and shear modules have been calculated by using the multiple scattering approach (MSA). These calculations allow us to estimate both frequency and loading dependencies of phase velocities of elastic waves in a wide range of frequencies and concentrations. Experimental verification of the MSA model was performed by comparing the calculated [52] and measured [53] phase velocities of the longitudinal wave in epoxy composites under high levels of filling (10 and 15%) with spherical particles having a diameter of a few dozen of μm. In this case, the agreement of the simulation results with the experimental observations was quite good. However, there was a lack of information about the volume of the interphase layer in such composites.

Another area of theoretical consideration is related to simulations of elastic constants performed in the self-consistent effective medium approximation [54]. In this case (sometimes referred to as the equivalent inclusion concept), the mechanical parameters of a three-phase composite system are quantitatively analyzed implying the particles, randomly dispersed in a matrix, can be considered as two-phase objects, which are composed of a nucleus particle surrounded by a layer of another phase. Three-phase composites filled with spherical [11], cylindrical [56], and ellipsoidal [54] particles have been successfully treated using this concept. The applicability of the effective medium approximation model for the three-phase composite system was first demonstrated by Dunn and Ledbetter [11], who measured the phase velocities of elastic waves in a three-phase ceramic composite. It was composed of an aluminum alloy reinforced by two-phase Al_2O_3/mullite spherical particles, which were 30, 45, or 100 μm in diameter. The embedded particle consisted of a short Al_2O_3 rod with an average aspect ratio (length to width) of approximately four dispersed in the mullite (a silicate mineral). The volume fraction of Al_2O_3 in the mullite was 20%, whereas the volume fraction of the particles in the Al alloy varied from 14.5 to

24%. The results from this theoretical analysis are in excellent agreement with the measured elastic constants.

Recently, the three-phase model was extended to polymeric composites with solid two-phase ellipsoidal particles [55]. However, as far as we are aware, an experimental verification of the three-phase model in graphene-based nanocomposites has not yet been reported.

Therefore, despite the important role of the filler/matrix interphase, the corresponding theory has not been fully developed, and an apparent discrepancy between previous investigations and theoretical treatments is found. This demonstrates the need to develop a further understanding of the contributions of the filler/matrix interphase area and changes in the composite stiffness regarding the material reinforcement effect.

4.4 Measurement Techniques

Acoustic waves are successfully applied to the evaluation of complex elastic modulus in a broad range of polymer composite [57–60]. Many aspects of measurement techniques can be found in [61–63]. The acoustic wave velocity measurement techniques are split into three main categories: wave propagation methods, resonance methods, and forced-vibration nonresonance methods. The resonance and forced-vibration techniques are designed to measure directly the complex Young's and shear moduli, and the acoustic properties of the polymer are calculated from this data. A comprehensive review of measurement techniques and data for the complex Young's modulus for a range of polymers is presented elsewhere [64]. In [65] was considered wave propagation methods of measurement for ultrasonic velocity and attenuation. These methods are mostly divided into the pulse-echo method and continuous wave transmission techniques (phase adjustment or π–point method). This chapter is restricted to a discussion of a modeling technique that could be useful for a better understanding of wave propagation methods when two piezoelectric transducers are connected with opposite sides of the specimen (see Fig. 4.6a).

4.4.1 Wave-Propagation Methods

Let's consider a piezoceramic disk with the thickness of L and poled in the direction z, as shown in Fig. 4.5a. The left and right sides of the plate were metalized.

In the general case, the vibration of the piezoelectric body is described by the following equations [66]: The equation of motion

$$\rho \frac{\partial^2 u_i}{\partial t^2} = \frac{\partial T_{ij}}{\partial x_j}. \tag{4.28}$$

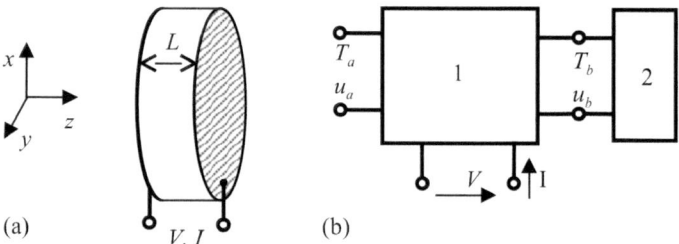

Fig. 4.5 a Schematics of the piezoelectric transducer with thickness L, V—applied rf voltage;
b six-pole network model (1) for the transducer with an acoustic load (2)

Poisson's equation

$$\frac{\partial D_i}{\partial x_i} = 0. \tag{4.29}$$

The stress tensor

$$T_{ij} = c^E_{ijkl} S_{kl} - e_{mij} E_m. \tag{4.30}$$

The electric displacement vector

$$D_i = e_{ikl} S_{kl} + \varepsilon^S_{ij} E_j. \tag{4.31}$$

The strain

$$S_{kl} = \frac{1}{2}\left(\frac{\partial u_k}{\partial x_l} + \frac{\partial u_l}{\partial x_k}\right), \tag{4.32}$$

where x_i is the coordinate, u_i is the displacement components, t is time, ρ is the mass density, c^E_{ijkl} are the elastic moduli of the medium under constant electric field, e_{ikl} are its piezoelectric coefficients, ε^S_{ij} the permittivity tensor under constant strain. All indices vary from 1 to 3 ($i = 1, 2, 3$) and repeated indices are summed. Electric field

$$E_i = -\frac{\partial \varphi}{\partial x_i} \tag{4.33}$$

and φ is a potential.

Given that the thickness of the transducer was much smaller than the diameter of the disk, we could consider a one-dimensional vibration body. Assuming the wave is a harmonic plane wave traveling in the z direction, the displacement components are defined by

$$u_i = u_{0i} e^{j(\omega t - kz)}, \tag{4.34}$$

where k is the wave number. Using complex wave numbers $k = k' - jk''$ allows us to take into account the losses in the transducer.

Substituting (4.34) with $i = 1, 2, 3$ into (4.28)–(4.31) yields a few solutions for one longitudinal and two transverse waves. For longitudinal waves, one can use the expression for the longitudinal speed of sound, and for transverse waves, the speed $v_t = \sqrt{c_{44}/\rho}$.

In our particular case, only the longitudinal wave with the velocity of $v_l = \sqrt{c_{33}/\rho}$ is relevant to the transducer shown in Fig. 4.5a. In this case, there would be two longitudinal waves counterpropagating in the z and $-z$ directions. Omitting the exponential term $e^{j\omega t}$ and taking into account that only the $E_3 (z)$ component of the electric field strength is not equal to zero, one equates:

$$u_1 = a_1 e^{-jkz} + b_1 e^{jkz}, \tag{4.35}$$

$$\varphi = -\int E_3 dz. \tag{4.36}$$

The electrical current through the transducer is given by $I = A\frac{\partial D_3}{\partial t} = j\omega A D_3$, with A being the cross section of the transducer. Dividing the transducer into a stack of thin layers, the displacement in the i-th layer takes the form:

$$u_i = a_i e^{-jk(z-z_i)} + b_i e^{jk(z-z_i)}, \tag{4.37}$$

which results in the stress in the i-th layer:

$$T_i = jk_i c_i \left(-a_i e^{-jk(z-z_i)} + b_i e^{jk(z-z_i)}\right) - e_i E_3, \tag{4.38}$$

where $c_i = c_{33}^E + e_{33}^2/\varepsilon_{33}^S$ and $e_i = e_{33}$ for the i-th layer. The voltage drop across the i-th layer is then given by:

$$V_i = -\int_{z_i}^{z_{i+1}} E_3 dz = \frac{e_i}{\varepsilon_i} \left(a_i(e^{-jk_i L_i} - 1) + b_i(e^{jk_i L_i} - 1)\right) + j\frac{I_i L_i}{\omega A \varepsilon_i}, \tag{4.39}$$

where ε_i and I_i are the permittivity and the current in the i-th layer, respectively, and $L_i = z_{i+1} - z_i$ is the thickness of the i-th layer. One can then employ the impedance matrix approach to describe the transducer [62, 66, 67]. For this, the boundary conditions for the i-th layer take the form:

$$u_{ai} = u_{i-1}(z_i) = u_i(z_i), \ u_{bi} = u_{i-1}(z_{i+1}) = u_i(z_{i+1}) \tag{4.40}$$

$$T_{ai} = T_{i-1}(z_i) = T_i(z_i), \ T_{bi} = T_{i-1}(z_{i+1}) = T_i(z_{i+1}) \tag{4.41}$$

Using Eqs. (4.40) and (4.41), one writes the following system of three linear equations:

$$
\begin{bmatrix} T_{ai} \\ T_{bi} \\ V_i \end{bmatrix} = \begin{bmatrix} -\dfrac{c_i k_i}{\tan(k_i L_i)} & \dfrac{c_i k_i}{\sin(k_i L_i)} & \dfrac{je_i}{\omega A \varepsilon_i} \\ -\dfrac{c_i k_i}{\sin(k_i L_i)} & \dfrac{c_i k_i}{\tan(k_i L_i)} & \dfrac{je_i}{\omega A \varepsilon_i} \\ \dfrac{e_i}{\varepsilon_i} & -\dfrac{e_i}{\varepsilon_i} & -\dfrac{jL_i}{\omega A \varepsilon_i} \end{bmatrix} \cdot \begin{bmatrix} u_{ai} \\ u_{bi} \\ I_i \end{bmatrix} \qquad (4.42)
$$

where $e_i = e_{33}$ and $\varepsilon_i = \varepsilon_{33}^S$.

The square 3×3 matrix was a three-port impedance matrix. Since each of the three ports had two transmission lines, a six-pole network was formed, as shown in Fig. 4.5b. The network had two acoustic ports, which are marked as T_{ai}, u_{ai} and T_{bi}, u_{bi} in Fig. 4.5b, and one electrical port V_i, I_i. In the case of non-piezoelectric vibrating disks, the electrical port was absent in the network, so one would consider only two acoustic ports shown in Fig. 4.5b. Hence, a solid layer of finite thickness is a two-port element forming a four-pole network. With increasing layer thickness to infinity, one would attain a two-pole network or a one-port element, which acts as the absorber of ultrasonic waves. At the absorber edge, the amplitude of the forward wave traveling from infinity to the port was equal to zero such that the stress and displacement could be related through some coefficient Z as $T_b = Z_b u_b$. The coefficient Z is equal to acoustic impedance multiplied by $j\omega$. If $Z = 0$, then it means that the surface is free from acoustic load and the stress on the surface will be zero.

Let's consider the main combinations of inclusion of 3-, 2- and 1-pole elements. The 2-pole element can be both purely acoustic, i.e., it will be a layer of material with a certain thickness, and electro-acoustic, i.e., a transducer acoustically loaded with some acoustic resistance.

The acoustic two-port element $q_{2\times2}$ connected to a one-port element with the impedance Z_b at the right-hand side, as shown in Fig. 4.5b, becomes a two-pole network with:

$$
Z_a = q_{1,1} + \frac{q_{1,2} q_{2,1}}{Z_b - q_{2,2}} \qquad (4.43)
$$

Instead, the acoustic two-port element $q_{2\times2}$ connected to a one-port element with the impedance Z_a at the left-hand side has

$$
Z_b = q_{2,2} + \frac{q_{1,2} q_{2,1}}{Z_a - q_{1,1}} \qquad (4.44)
$$

The electro-acoustic three-port element $p_{3\times3}$ connected to a one-port element with the impedance Z_b at the right-hand side, as shown in Fig. 4.5b, becomes a two-port electro-acoustic element $r_{2\times2}$ with components:

$$
rb_{1,1} = p_{1,1} + \frac{p_{2,1} p_{1,2}}{Z_b - p_{2,2}}, rb_{1,2} = p_{1,3} + \frac{p_{2,3} p_{1,2}}{Z_b - p_{2,2}}, \qquad (4.45)
$$

$$
rb_{2,1} = p_{3,1} + \frac{p_{2,1} p_{3,2}}{Z_b - p_{2,2}}, rb_{2,2} = p_{3,3} + \frac{p_{2,3} p_{3,2}}{Z_b - p_{2,2}}. \qquad (4.46)
$$

Instead, the electro-acoustic three-port element $p_{3\times3}$ connected to a one-port element with the impedance Z_a at the left-hand side has

$$ra_{1,1} = p_{2,2} + \frac{p_{1,2}p_{2,1}}{Z_a - p_{1,1}}, ra_{1,2} = p_{2,3} + \frac{p_{1,3}p_{2,1}}{Z_a - p_{1,1}}, \tag{4.47}$$

$$ra_{2,1} = p_{3,2} + \frac{p_{1,2}p_{3,1}}{Z_a - p_{1,1}}, ra_{2,2} = p_{3,3} + \frac{p_{1,3}p_{3,1}}{Z_a - p_{1,1}}. \tag{4.48}$$

If the three-port element $p_{3\times3}$ is acoustically loaded by one-port elements with the impedances Z_a and Z_b at the left- and right-hand sides, respectively, the resistance of the appropriately loaded transducer can be found as

$$R = p_{3,3} + \frac{p_{3,1}\left(p_{1,2}p_{2,3} + p_{1,3}\left(Z_b - p_{2,2}\right)\right) + p_{3,2}\left(p_{1,3}p_{2,1} + p_{2,3}\left(Z_a - p_{1,1}\right)\right)}{Z_a Z_b - Z_a p_{2,2} - Z_b p_{1,1} + p_{1,1}p_{2,2} - p_{1,2}p_{2,1}}$$
$$\tag{4.49}$$

Therefore, using Eqs. (4.45) and (4.44), the transducer with any number of loading layers can be described by the two-pole network with the resistance given by Eq. (4.49).

We can also write down the expression for the transfer function between the input and output voltages, if two converters are used between which the sample is located (Fig. 4.6c)

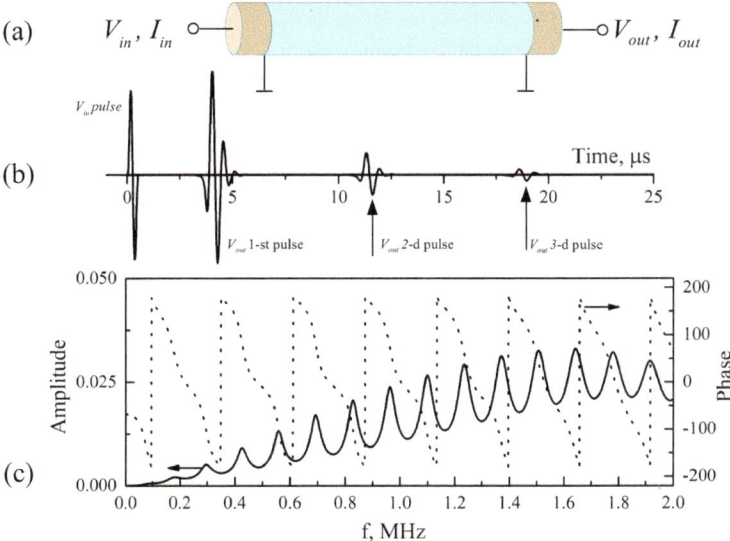

Fig. 4.6 Schematics of a combination where two converters are used between which the sample is located (**a**). Response to impulse short excitation (**b**). Transfer function in frequency domain (**c**)

$$H = \frac{t_{1,2}r_{2,1}}{t_{1,1}t_{2,2} - rb_{1,2}t_{2,1} - t_{2,2}rb_{1,1}}, \tag{4.50}$$

where components $rb_{2\times2}$ are taken from Eqs. (4.45) and (4.46). The components $t_{2\times2}$ are calculated by expressions

$$t_{1,1} = q_{2,2} - \frac{q_{2,1}q_{1,2}}{q_{1,1} - r_{1,1}}, t_{1,2} = \frac{ra_{1,2}q_{2,1}}{q_{1,1} - r_{1,1}}, \tag{4.51}$$

$$t_{2,1} = -\frac{ra_{2,1}q_{1,2}}{q_{1,1} - r_{1,1}}, t_{2,2} = ra_{2,2} + \frac{ra_{2,1}ra_{1,2}}{q_{1,1} - r_{1,1}}, \tag{4.52}$$

where $q_{2\times2}$ is the acoustic two-port element (plastic rod).

Using expression (4.50), we can find not only the transfer function of the two-port delay line, but also its response to impulse excitation by voltage. To do this, we will use the fast Fourier transform (FFT) and the concept of convolution of two functions in the time domain. That is, the time input pulse $p(t)$ is converted to its spectrum $\hat{p}_n$ through the use of the forward FFT. The transformed solution is then obtained by evaluating the product $\hat{u}_n = \hat{p}_n H(f_n)$ at each frequency. This is finally reconstructed in the time domain by use of the inverse FFT. Details can be found elsewhere [68].

The transfer function on Fig. 4.6c was calculated with parameters typical for epoxy resin [58]: damping factor $\eta = 0.028$ (damping factor η is equal to the ratio of the imaginary part E'' to the real part E' of a complex elastic modulus, $E^* = E - iE''$), longitudinal wave velocity 2750 m/s and density 1.39 g/cm^3. Delay line length $L = 10$ mm and damping factor $\eta = 1$ for both transducers. With that high damping factor, one has to construct wideband transducers because a short pulse is applied to the delay line.

Another type of wave propagation technique is a continuous wave transmission technique also called ultrasound phase spectroscopy (UPS). In UPS, continuous and sinusoidal elastic waves are passed through the sample, and the phase shift is measured as a function of frequency [69–72]. UPS is an optimal technique to determine sound velocities in samples exhibiting large attenuations. The base idea of the UPS can be explained by Fig. 4.6c where the phase shift of the transfer function periodically changes with the frequency.

Different pairs of transducers are typically used to excite longitudinal and shear waves separately. In every case, frequency dependencies of the group delay time T_g and the group velocity v_g can be determined by using the well-known relations

$$T_g(\omega) = \frac{d\Phi(\omega)}{d\omega}, \tag{4.53}$$

$$v_g(\omega) = \frac{d\omega}{dk} = \frac{L_S}{T_g(\omega)}, \tag{4.54}$$

$$k\left(\omega\right) = \frac{\Phi\left(\omega\right)}{L_S},\tag{4.55}$$

where $k\left(\omega\right)$ is the wave number and L_S is the sample's thickness. Then the phase velocity $v\left(\omega\right)$ can be evaluated from [73]

$$\frac{1}{v_g(\omega)} = \frac{1}{v(\omega)} - \frac{\omega}{v^2(\omega)}\frac{dv(\omega)}{d\omega}.\tag{4.56}$$

In the case of negligible dispersion $dv(\omega)/d\omega \approx 0$ so that (4.56) is reduced to $v\left(\omega\right) \approx v_g\left(\omega\right)$.

To exclude an overestimate of T_g due to wave propagation through the two transducers, both v_L and v_S measurements can be made with two samples of different lengths, say, L_1 and $L_2 = L_1/2$. In this case, the relation

$$v_g\left(\omega\right) = \frac{L_1 - L_2}{T_{g1}\left(\omega\right) - T_{g2}\left(\omega\right)}\tag{4.57}$$

can be used instead of (4.56) to calculate $v_g(\omega)$. Here, both $T_{g1}(\omega)$ and $T_{g2}(\omega)$ are determined as inclination angle tangents of straight lines originating from the least-squares-approximated experimental $\Phi\left(\omega\right)$ dependencies.

When the values of v_L and v_S are measured, the Lame constants λ and μ are evaluated via v_L and v_S and calculated value of the sample density ρ by [66]

$$\rho v_L^2 = \lambda + 2\mu, \quad \rho v_S^2 = \mu.\tag{4.58}$$

Finally, a set of mechanical parameters including Young's modulus E, the compression modulus K, and the Poisson's ratio v can be calculated via the Lame constants λ and μ by using the correspondent expressions which are valid for rod-shaped samples [66]

$$v = \frac{\lambda}{2(\lambda + \mu)}, \quad E = \mu\frac{3\lambda + 2\mu}{\lambda + \mu}, \quad K = \lambda + 2\mu/3.\tag{4.59}$$

4.4.2 Resonance Techniques

The resonance test method covers the measurement of the fundamental transverse, longitudinal, and torsional frequencies of isotropic and anisotropic materials. When an external force (driver) acts on the specimen the amplitude of oscillations is measured in one way or another. That is, this method requires the correct shape specimens to excite free oscillations in them. These measured resonant frequencies are used to calculate dynamic elastic moduli for necessary orientations. The method was described in the ASTM C747 standard [74] and does not specify the way of action on

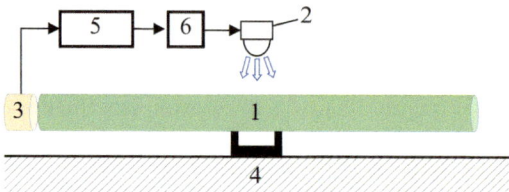

Fig. 4.7 Schematic diagram of the measurement setup. Modulated light absorption forces rod vibrations, which are sensed by a microphone through an air pressure wave. 1—rod-shaped samples, 2—light-emitting diode (LED), 3—microphone, 4—support, 5—Lock-in amplifier, 6—LED driver

the sample and the vibration amplitude sensor, but only gives general recommendations for them. The most important of which is to provide a minimal contact pressure on the sample from the driver and vibration sensor, so as not to change the frequency of the sample vibrations. Here, we describe one of the possible setups for measuring the natural oscillation frequency of epoxy-graphene nanocomposite samples [75]. Some more works that also contain an extensive list of references on the resonance test method can be found in Jarzynski et al. [64] and Nadtochiy et al. [76].

The scheme of the experimental setup is shown in Fig. 4.7. The measurement of the sound velocity in rods was carried out using the fundamental resonant frequency of a slender rod of circular cross section. We used the thermoelastic method to excite acoustic vibrations. The light-emitting diode (LED) light strikes the middle part of the rod and the absorbed radiation is transformed into heat, which depends on the absorption coefficient and the incident light intensity. The sample was therefore mechanically unloaded, which improved the accuracy of measuring its natural frequencies. Tuning the LED modulation frequency, resonance conditions are easily achieved at the natural frequencies of the rod. The released modulated heat appears as acoustic and thermal waves in the sample generated due to localized transient heating and expansion. The end of the sample has the maximum amplitude of oscillations and is the source of sound. The vibration amplitude was sensed by a microphone through an air pressure wave. For a frequency range less than 20 kHz, a conventional electret microphone can be used, and for frequencies higher than 20 kHz, other types of air-coupled [77, 78] ultrasonic transducers or non-contact remote optical detection of ultrasound [79].

For rods and bars, the dynamic modulus of elasticity can be calculated from the fundamental longitudinal frequency, weight, and dimensions of the test sample as [74] $E = 4f^2L^2\rho$, where f is the frequency of fundamental longitudinal mode of vibration, L is the length of the sample, and ρ is its density.

References

1. H.F. Brinson, L.C. Brinson, *Polymer Engineering Science and Viscoelasticity: An Introduction* (Springer, 2015)
2. H.J. McSkimin, Ultrasonic methods for measuring the mechanical properties of liquids and solids, in *Physical Acoustics*, vol. 1A, ed. by W.P. Mason (Academic Press, New York, 1964)
3. H.M. Nussenzveig, *Causality and Dispersion Relations*, 1st edn. (Academic, New York, 1972)
4. M. O'Donnell, E.T. Jaynes, J.G. Miller, General relationships between ultrasonic attenuation and dispersion. J. Acoust. Soc. Am. **63**, 1935–1937 (1978)
5. M. O'Donnell, E.T. Jaynes, J.G. Miller, Kramers–Kronig relationship between ultrasonic attenuation and phase velocity. J. Acoust. Soc. Am. **69**, 696–701 (1981)
6. Y. Wang, R.E. Challis, A.P.Y. Phang, M.E. Unwin, Bulk shear wave propagation in an epoxy: attenuation and phase velocity over five decades of frequency. IEEE Trans. Ultrason. Ferroelectr. Freq. Control **56**, 2504–2513 (2009)
7. A.N. Netravali, K.L. Mittal (eds.), *Interface/Interphase in Polymer Nanocomposites* (Wiley, Hoboken, NJ, 2017)
8. T. Mori, K. Tanaka, Average stress in matrix and average elastic energy of materials with misfitting inclusions. Acta Metall. **21**, 571–574 (1973)
9. G.M. Odegard, T.C. Clancy, T.S. Gates, Modeling of the mechanical properties of nanoparticle/polymer composites. Polymer **46**, 553–562 (2005)
10. T. Mura, *Micromechanics of Defects in Solids*, 2nd edn. (Martinus Nijhoff, Dordrecht, 1987)
11. M.L. Dunn, H. Ledbetter, Elastic moduli of composites reinforced by multiphase particles. Appl. Mech. **62**, 1023–1028 (1995)
12. C. Lee, X. Wei, J.W. Kysar, J. Hone, Measurement of the elastic properties and intrinsic strength of monolayer graphene. Science **321**, 385–388 (2008)
13. J.R. Potts, D.R. Dreyer, C.W. Bielawski, R.S. Ruoff, Graphene-based polymer nanocomposites. Polymer **52**, 5–25 (2011)
14. B.M. Gorelov, A. Gorb, A. Nadtochiy, D. Starokadomsky, V. Kuryliuk, N. Sigareva, S.V. Shulga, V.M. Ogenko, O. Korotchenkov, O. Polovina, Epoxy filled with bare and oxidized multi-layered graphene nanoplatelets: a comparative study of filler loading impact on thermal properties. J. Mater. Sci. **54**, 9247–9266 (2019)
15. A.B. Nadtochiy, A.M. Gorb, B.M. Gorelov, O.I. Polovina, O. Korotchenkov, V. Schlosser, Model approach to thermal conductivity in hybrid graphene-polymer nanocomposites. Molecules **28**, 7343 (2023)
16. B.M. Gorelov, O.V. Mischanchuk, N.V. Sigareva, S.V. Shulga, A.M. Gorb, O.I. Polovina, V.O. Yukhymchuk, Structural and dipole-relaxation processes in epoxy-multilayer graphene composites with low filler content. Polymers **13**, 3360 (2021)
17. R. Hashemi, On the overall viscoelastic behavior of graphene/polymer nanocomposites with imperfect interface. Int. J. Eng. Sci. **105**, 38–55 (2016)
18. X. Huang, Z. Chen, Y. Lin, H. Bao, G. Wu, P. Jiang, Y. Mai, Thermal conductivity of graphene-based polymer nanocomposites. Mater. Sci. Eng. R. Rep. **142**, 100577 (2020)
19. A. Tarhini, Graphene-based polymer composite films with enhanced mechanical properties and ultra-high in-plane thermal conductivity. Compos. Sci. Technol. **184**, 107797 (2019)
20. D.G. Papageorgiou, I.A. Kinloch, R.J. Young, Mechanical properties of graphene and graphene-based nanocomposites. Prog. Mater. Sci. **90**, 75–127 (2017)
21. R. Rafiee, A. Eskandariyun, Predicting Young's modulus of agglomerated graphene/polymer using multi-scale modeling. Compos. Struct. **245**, 112324 (2020)
22. J. Wang, F. Song, Y. Ding, M. Shao, The incorporation of graphene to enhance mechanical properties of polypropylene self-reinforced polymer composites. Mater. Des. **195**, 109073 (2020)
23. D.R. Paul, L.M. Robeson, Polymer nanotechnology: nanocomposites. Polymer **49**, 3187–3204 (2008)

24. J. Jančář, J.F. Douglas, F.W. Starr, S.K. Kumar, P. Cassagnau, A.J. Lesser, S.S. Sternstein, M.J. Buehler, Current issues in research on structure-property relationships in polymer nanocomposites. Polymer **51**, 3321–3343 (2010)
25. A. Zhu, S.S. Sternstein, Nonlinear viscoelasticity of nanofilled polymers: interfaces, chain statistics and properties recovery kinetics. Compos. Sci. Technol. **63**, 1113–1126 (2003)
26. Q. Zeng, A. Yu, G.Q. Lu, Multiscale modeling and simulation of polymer nanocomposites. Prog. Polym. Sci. **33**, 191–269 (2008)
27. A. Montazeri, R. Naghdabadi, Investigation of the interphase effects on the mechanical behavior of carbon nanotube polymer composites by multiscale modeling. J. Appl. Polym. Sci. **117**, 361–367 (2010)
28. S. Yang, M. Cho, Scale bridging method to characterize mechanical properties of nanoparticle/polymer nanocomposites. Appl. Phys. Lett. **93** (2008)
29. S. Yu, S. Yang, M. Cho, Multi-scale modeling of cross-linked epoxy nanocomposites. Polymer **50**, 945–952 (2009)
30. A. Gooneie, S. Schuschnigg, C. Holzer, A review of multiscale computational methods in polymeric materials. Polymers **9**, 16 (2017)
31. X. Wu, A. Aramoon, J.A. El-Awady, Hierarchical multiscale approach for modeling the deformation and failure of Epoxy-Based polymer matrix composites. J. Phys. Chem. B. **124**, 11928–11938 (2020)
32. R. Rafiee, R. Shahzadi, Predicting mechanical properties of nanoclay/polymer composites using stochastic approach. Compos. B. Eng. **152**, 31–42 (2018)
33. M. Zahedi, R. Malekimoghadam, R. Rafiee, U. Icardi, A study on fracture behavior of semi-elliptical 3D crack in clay-polymer nanocomposites considering interfacial debonding. Eng. Fract. Mech. **209**, 245–259 (2019)
34. R. Rafiee, H. Zehtabzadeh, Predicting the strength of carbon nanotube reinforced polymers using stochastic bottom-up modeling. Appl. Phys. A **126** (2020)
35. R. Rafiee, M. Sahraei, Characterizing delamination toughness of laminated composites containing carbon nanotubes: experimental study and stochastic multi-scale modeling. Compos. Sci. Technol. **201**, 108487 (2021)
36. R. Rafiee, A. Ghorbanhosseini, Predicting mechanical properties of fuzzy fiber reinforced composites: radially grown carbon nanotubes on the carbon fiber. Int. J. Mech. Mater. Des. **14**, 37–50 (2016)
37. R. Rafiee, A. Ghorbanhosseini, Stochastic multi-scale modeling of randomly grown CNTs on carbon fiber. Mech. Mater. **106**, 1–7 (2017)
38. R. Rafiee, A. Eskandariyun, Estimating Young's modulus of graphene/polymer composites using stochastic multi-scale modeling. Compos. B. Eng. **173**, 106842 (2019)
39. A.S. Sarvestani, C.R. Picu, Network model for the viscoelastic behavior of polymer nanocomposites. Polymer **45**, 7779–7790 (2004)
40. C.M. Sahagun, K.M. Knauer, S.E. Morgan, Molecular network development and evolution of nanoscale morphology in an epoxy-amine thermoset polymer. J. Appl. Polym. Sci. **126**, 1394–1405 (2012)
41. S. Morsch, Y. Liu, P. Greensmith, S.B. Lyon, S. Gibbon, Molecularly controlled epoxy network nanostructures. Polymer **108**, 146–153 (2017)
42. U. Yamamoto, K.S. Schweizer, Theory of nanoparticle diffusion in unentangled and entangled polymer melts. J. Chem. Phys. **135** (2011)
43. T. Tanaka, M. Kozako, N. Fuse, Y. Ohki, Proposal of a multi-core model for polymer nanocomposite dielectrics. IEEE Trans. Dielectr. Electr. Insul. **12**, 669–681 (2005)
44. R. Atif, I. Shyha, F. Inam, Modeling and experimentation of multi-layered nanostructured graphene-epoxy nanocomposites for enhanced thermal and mechanical properties. J. Compos. Mater. **51**, 209–220 (2016)
45. M. Li, H. Zhou, Y. Zhang, Y. Liao, H. Zhou, Effect of defects on thermal conductivity of graphene/epoxy nanocomposites. Carbon **130**, 295–303 (2018)
46. S.I. Kundalwal, S.A. Meguid, G.J. Weng, Strain gradient polarization in graphene. Carbon **117**, 462–472 (2017)

47. J. Chung, M. Munz, H. Stürm, Stiffness variation in the interphase of amine-cured epoxy adjacent to copper microstructures. Surf. Interface Anal. **39**, 624–633 (2007)
48. C. Wehlack, W. Possart, J. Krüger, U. Müller, Epoxy and Polyurethane networks in thin films on metals—formation, structure, properties. Soft Mater. **5**, 87–134 (2007)
49. K. Farah, F. Leroy, F. Müller-Plathe, M. Böhm, Interphase formation during curing: reactive coarse grained molecular dynamics simulations. J. Phys. Chem. C **115**, 16451–16460 (2011)
50. R.M. Christensen, K.H. Lo, Solutions for effective shear properties in three phase sphere and cylinder models. J. Mech. Phys. Solids **27**, 315–330 (1979)
51. C.N. Layman, N.S. Murthy, R. Yang, J. Wu, The interaction of ultrasound with particulate composites. J. Acoust. Soc. Am. **119**, 1449–1456 (2006)
52. R. Yang, A dynamic generalized self-consistent model for wave propagation in particulate composites. Appl. Mech. **70**, 575–582 (2003)
53. V.K. Kinra, M.S. Petraitis, S.K. Datta, Ultrasonic wave propagation in a random particulate composite. Int. J. Solids Struct. **16**, 301–312 (1980)
54. M. Hori, S. Nemat-Nasser, Double-inclusion model and overall moduli of multi-phase composites. Mech. Mater. **14**, 189–206 (1993)
55. P. Lü, Y.W. Leong, P.K. Pallathadka, C. He, Effective moduli of nanoparticle reinforced composites considering interphase effect by extended double-inclusion model – Theory and explicit expressions. Int. J. Eng. Sci. **73**, 33–55 (2013)
56. Y. Huang, K. Hu, X. Wei, A. Chandra, A generalized self-consistent mechanics method for composite materials with multiphase inclusions. J. Mech. Phys. Solids **42**, 491–504 (1994)
57. K. Ono, A comprehensive report on ultrasonic attenuation of engineering materials, including metals, ceramics, polymers, fiber-reinforced composites, wood, and rocks. Appl. Sci. **10**, 2230 (2020)
58. K. Ono, Dynamic viscosity and transverse ultrasonic attenuation of engineering materials. Appl. Sci. **10**, 5265 (2020)
59. K. Ono, Ultrasonic attenuation of ceramic and inorganic materials using the through-transmission method. Appl. Sci. **12**, 13026 (2022)
60. İ. Oral, Determination of elastic constants of epoxy resin/biochar composites by ultrasonic pulse echo overlap method. Polym. Compos. **37**, 2907–2915 (2015)
61. L.C. Lynnworth, *Ultrasonic Measurements for Process Control: Theory Techniques, Applications* (1989)
62. L.W. Schmerr, J.-S. Song, *Ultrasonic Nondestructive Evaluation Systems: Models and Measurements* (Springer, 2007)
63. B. Lüthi, *Physical Acoustics in the Solid State* (Springer, 2005)
64. J. Jarzynski, E. Balizer, J.J. Fedderly, G. Lee, Acoustic properties. Encyclopedia of Polymer Science and Technology (2003)
65. E.P. Papadakis, Ultrasonic velocity and attenuation: measurement methods with scientific and industrial applications, in *Physical Acoustics*, vol. XII, ed. by W.P. Mason (1976), pp. 277–374
66. D. Royer, E. Dieulesaint, *Elastic Waves in Solids I: Free and Guided Propagation* (Springer, 1999)
67. B.A. Auld, *Acoustic Fields and Waves in Solids* (Wiley-Interscience, 1973)
68. J.F. Doyle, *Wave Propagation in Structures: An FFT-Based Spectral Analysis Methodology* (Springer, 1989)
69. E.P. Papadakis, The measurement of small changes in ultrasonic velocity and attenuation. Crit. Rev. Solid State Mater. Sci. **3**, 373–418 (1973)
70. A. Wanner, Elastic modulus measurements of extremely porous ceramic materials by ultrasonic phase spectroscopy. Mater. Sci. Eng. A **248**, 35–43 (1998)
71. S. Roy, J.-M. Gebert, G. Stasiuk, R. Piat, K.A. Weidenmann, A. Wanner, Complete determination of elastic moduli of interpenetrating metal/ceramic composites using ultrasonic techniques and micromechanical modelling. Mater. Sci. Eng. A **528**, 8226–8235 (2011)
72. C.S. Ting, Measurement of ultrasonic dispersion by phase comparison of continuous harmonic waves. J. Acoust. Soc. Am. **64**, 852–857 (1978)
73. R.B. Lindsay, *Mechanical Radiation* (McGraw-Hill, New York, 1960)

74. ASTM C747-93(2010): Standard Test Method for Moduli of Elasticity and Fundamental Frequencies of Carbon and Graphite Materials by Sonic Resonance (2010) ASTM International, West Conshohocken, PA. www.astm.org
75. A. Nadtochiy, B.M. Gorelov, O. Polovina, S.V. Shul'ga, O. Korotchenkov, Probing matrix/filler interphase with ultrasonic waves. J. Mater. Sci. **56**, 14047–14069 (2021)
76. A. Nadtochiy, O.K. Suwal, D.-S. Kim, O. Korotchenkov, Revealing CdSe quantum dots plasmonics confined in Au nanotrenches by thermoacoustic spectroscopy. ACS Appl. Opt. Mater. **1**, 1272–1280 (2023)
77. D.E. Chimenti, Review of air-coupled ultrasonic materials characterization. Ultrasonics **54**, 1804–1816 (2014)
78. T.G. Alvarez-Arenas, J. Camacho, Air-coupled and resonant pulse-echo ultrasonic technique. Sensors **19**, 2221 (2019)
79. J. Spytek, L. Ambroziński, I. Pelivanov, Non-contact detection of ultrasound with light – Review of recent progress. Photoacoustics **29**, 100440 (2023)